R. J. Goult   P. A. Sherar (Eds.)

Improving the Performance of Neutral File Data Transfers

Springer-Verlag

Berlin Heidelberg New York London
Paris Tokyo Hong Kong Barcelona

Volume Editors

Raymond J. Goult
Peter A. Sherar
Department of Applied Computing and Mathematics
Cranfield Institute of Technology
Cranfield, Bedford MK43 0AL, UK

Science TS
155.6
I5
1990

ESPRIT Project 322: CAD Interfaces (CAD*I) belongs to the Research and Development area "Computer-Aided Design and Engineering (CAD / CAE)" within the Subprogramme 5 "Computer-Integrated Manufacturing (CIM)" of the ESPRIT Programme (European Strategic Programme for Research and Development in Information Technology) supported by the European Communities.

ESPRIT Project 322 has been established to define the most important interfaces in CAD / CAM systems for data exchange, data base, finite element analysis, experimental analysis, and advanced modelling. The definitions of these interfaces are being elaborated in harmony with international standardization efforts in this field.

Partners in the project are:
Bayerische Motorenwerke AG / FRG · CISIGRAPH / France · Cranfield Institute of Technology / UK · Danmarks Tekniske Højskole / Denmark · Estudios y Realizaciones en Diseño Informatizado SA (ERDISA) / Spain · Gesellschaft für Strukturanalyse (GfS) mbH / FRG · Katholieke Universiteit Leuven / Belgium · Kernforschungszentrum Karlsruhe GmbH / FRG · Leuven Measurement and Systems / Belgium · NEH Consulting Engineers ApS / Denmark · Rutherford Appleton Laboratory / UK · Universität Karlsruhe / FRG.

CR Subject Classification (1987): J.6

ISBN 3-540-53427-X Springer-Verlag Berlin Heidelberg New York
ISBN 0-387-53427-X Springer-Verlag New York Berlin Heidelberg

Publication No. EUR 13184 EN of the
Commission of the European Communities,
Scientific and Technical Communication Unit,
Directorate-General Telecommunications, Information Industries and Innovation,
Luxembourg
Neither the Commission of the European Communities nor any person acting on behalf of the Commission is responsible for the use which might be made of the following information.

© ECSC – EEC – EAEC, Brussels – Luxembourg, 1990
Printed in Germany

Printing and Binding: Weihert-Druck GmbH, Darmstadt
2145/3140 – 543210 – Printed on acid-free paper

Research Reports ESPRIT

Subseries
Project 322 · CAD Interfaces (CAD*I)
Volume 6

T5

Subseries Editors:
I. Bey, Kernforschungszentrum Karlsruhe GmbH
J. Leuridan, Leuven Measurement and Systems

Edited in cooperation with
the Commission of the European Communities

Abstract

This is one of a series of books which present the results of ESPRIT project 322 'CAD Interfaces' (CAD*I). In this volume we concentrate on the work done for the transfer of data describing complex curves and surfaces. The book gives an introduction to some of the inherent problems in this type of communication and describes some algorithms and software tools which assist in this transfer. Requirements for an effective neutral file specification are described in the context of the development of the CAD*I neutral file specification and the proposed international standard STEP.

The book is primarily intended for readers who are interested in

- the problems encountered in the transfer of geometric data for curves and surfaces,

- the state of the art reached after completion of the project, and

- the relationship between the CAD*I project and the development of the international standard for external representation of product definition data STEP.

Contributors

R.J. Goult, Cranfield Institute of Technology

M.A. Lachance, Department of Mathematics, University of Michigan, Dearborn

H. Scheder, BMW AG, Munich

P.A. Sherar, Cranfield Institute of Technology

D. Trippner, BMW AG, Munich

Table of Contents

CAD*I Project Overview

During the past 25 years computers have been introduced in industry to perform technical tasks such as drafting, design, process planning, data acquisition, process control and quality assurance. Computer-based solutions, however, are still in most cases single isolated devices within a manufacturing plant.

Computer technology is evolving rapidly, and the life cycle of today's products and production methods is shortening, with continuously increasing requirements from customers, and a trend to market interrelations between companies at a national and international level. This forces a growing need for efficient storage, retrieval and exchange of information. Integration of information is urgent within companies to interconnect departments which used to work more or less on their own. On the other hand direct communication with outside customers, suppliers and partner institutions will often determine the position of an enterprise among its competitors. In this sense, Computer Integrated Manufacturing (CIM) is the key of today for the competitiveness of tomorrow. But the realization of a future-oriented CIM concept is not possible without powerful, widely accepted and standardized interfaces. They are the vital issue on the way to CIM. They will contribute to harmonizing data structures and information flows and will play a major role in open CIM systems. Standardized interfaces allow:

- Access to data produced and archived on computing equipment which is no longer in active use;
- Communication between hardware and software from different vendors;
- Paperless exchange of information.

ESPRIT Project 322 "CAD Interfaces" (CAD*I) started in 1984 is a five-year research and development programme on CAD interfaces with the aim of defining some missing interface specifications in the environment of computer aided design (CAD) systems for mechanical engineering. Parts design and CAD are the starting point in the design and manufacturing process, and can also be considered as a starting point for information generation and data exchange.

Based on the results and using the experiences of former national standardization initiatives like IGES, VDAFS or SET, the CAD*I project team aimed from early in the project to contribute to the first international standard for product data exchange, because only an internationally accepted standard interface will fulfill the requirements of European industry.

The standardization work in CAD data exchange at international level is performed through ISO/TC184/SC4 under the name STEP: Standard for the Exchange of Product Model Data. CAD*I has had a large influence on the STEP definitions especially for the exchange of geometry and shape information (curves, surfaces and solid models), the interface to Finite Element Analysis programmes and drafting information.

This report is one of a series of similar books which summarize the wealth of results achieved during the five years of ESPRIT Project CAD*I.

CAD Interfaces

The main results are:

- Vendor independent interface consisting of a neutral file specification and corresponding pre- and post-processors for many commercial CAD systems have been defined, developed and tested. The CAD*I specifications for geometry and shape representation (curves, surfaces and solids) are clearly visible in the first international draft proposal standard. The processors are in practical use in several European and national projects. European system vendors have begun to integrate these results into their products.

- A general standard specification of a neutral file for exchanging finite element data has been developed and implemented. Tests have been performed with the interface processors for several FEM packages available on the market. In addition CAD models were transferred to finite element systems using the CAD*I neutral file. The results of this work have already appeared on the European market.

- New and improved data acquisition and analytical procedures for dynamic structural analysis have been specified and tested on complex real structures. Also, powerful tools for the intelligent integration (link) of experimental and analytical results in structural design have been developed, tested and merged into software products now available on the market. These results are visible in recent commercial products.

- Some new methods have been developed to enhance the communication interface in CAD/CAE systems. Future users of this kind of system will be able to enter information to the systems by handsketching input or by technical terms from using design language instead of via formal geometrical descriptions. First implementations have been successful; they are based on levels of internal interfaces using a product model.

- A neutral database interface based on the CAD*I neutral file format has been developed to handle archiving and retrieval of product information in a database. A set of standard access routines has been written and tested with existing CAD systems and a widely used commercial relational database management system. The introduction of these results into marketable products is on the way.

- An information model for the description of technical drawings has been developed: the CAD*I drafting model. This information model represents the highest level of sophistication within the level concept of the drafting model of the STEP specification.

A total of about 150 person-years of research and development effort has been spent on the project. The CAD*I project involved 12 partners in 6 countries of the European Community.

As project manager since 1985 I would like to express my appreciation to the co-manager J. Leuridan and the fifty or more people working in and on the project for their engagement to reach the originally stated goals. In addition I would like to pay special tribute to Mrs. P. MacConaill and R. Zimmermann form the Commission of the European Communities and to the reviewers G. Enderle (+), E.A. Warman and H. Nowacki for their cooperative support. Special acknowledgement is due also to Mrs. U. Frey for running the administrative part of the project and for her contributions to forming the spirit of the CAD*I team.

I. Bey, CAD*I Project Manager

1 CAD/CAM DATA EXCHANGE IN THE INDUSTRIAL ENVIRONMENT - METHODOLOGY AND TOOLS

Principal authors: H. Scheder, D. Trippner

1.0   Introduction

In the last years the entire process of industrial development has become more and more computer-aided. In almost all areas and stages of product development, process planning and product manufacturing itself, CAx systems are in constant use to support the work in terms of productivity, quality and reliability. Due to the diversity of CAx applications, the systems used are generally well specified and appropriate to the relevant application. This results in the implementation of special sets of entities and structures in the systems databases.

In order to reach the goals usually set up with the introduction of computer-aided technologies, data describing technical products must be accessible to all applications requiring the data. This leads to strong interconnections between all parts of the enterprise involved in the development of a certain product.

Therefore the exchange of product definition data is a basic requirement for the profitable use of computer-aided technologies in industrial product development.

1.1   Experiences with CAD data transfer

To exchange product definition data between different CAx systems, neutral file interfaces are required. In addition to the national standards SET and VDAFS, the neutral file specification of IGES has been established as a national standard with international acceptance. Other specifications, like PDDI or CAD*I have reached some significance on the national or european level. STEP, the first international standard is still under development, the geometry part of the standard being at the prototype stage.

Today, IGES is the standard which is used by more companies in a wider range of industries than any other standard for CAD data exchange. Although the quality of implemented IGES processors has been improved during the last years, data exchange may not always be free of problems.

As shown in Figure 1, three classes of reasons for problems can still be recognised:

Problems

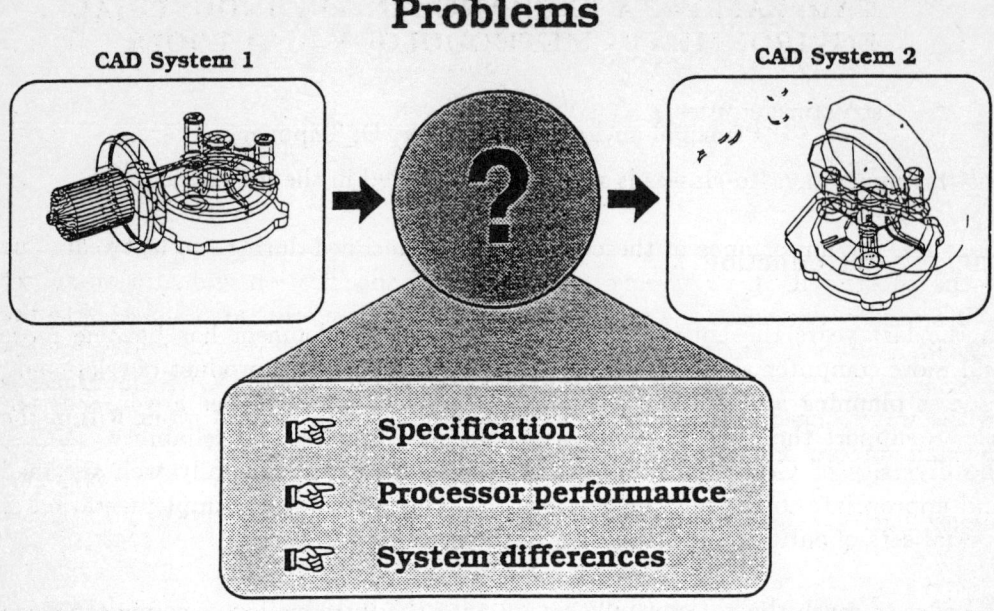

CAD System 1   CAD System 2

☞ **Specification**

☞ **Processor performance**

☞ **System differences**

Figure 1. Classes of problems in exchanging CAD/CAM data

* incomplete and ambiguous specification,
* performance of implemented pre- and post-processors and
* principle differences between CAD/CAM systems.

1.2   Problems caused by the specification

Although the IGES specification with its large set of entities covers a wide area of applications it does not always provide the right entity types for a special task. On the other hand redundant entity definitions exist in the specification. This can lead to data exchange errors even if both systems, the sending and the receiving system, are able to handle this kind of information.

For example, the entity definitions in the IGES specification are structured into the classes:

* geometry entities,

* annotation entities and

* structure entities,

but this assignment to classes is not strictly followed in the definition.

Of course the importance of these problems has declined during the last years due to the efforts put on new versions of the IGES specification and also on subset definitions. In particular the development of application orientated subsets like the VDAIS, a subset definition based on IGES, has lead to a more reliable and safer data exchange. For this reason the exchange of data via the VDAIS (IGES) interfaces is on the way to becoming a normal method of interaction in daily work within the German automotive industry.

Figure 2 gives a short impression of the subsets defined by the VDAIS .

1.3 Problems regarding the processor quality

Although the quality and performance of neutral file processors has increased in the last years, problems caused by different interpretations of the standard, bad implementation or simply programming errors still arise.

Fixing the errors can be a very complex task requiring specialist knowledge. Normally it is not possible to find bugs through the examination of IGES data, although the IGES format is a man readable ASCII format, because the file sizes are too big. So what can be done to find out the cause of the error? Which of the components involved in the data exchange is responsible for the error?

For the quick and reliable fixing of errors, for independent processor testing and for the support of daily data exchange, a set of basic data exchange software tools is imperative and needed to be developed. For effective use of these tools a special test methodology supplemented with relevant test data (see Figure 3) was necessary and was also introduced.

The test methodology provides different test procedures for independent examination of the several parts participating in the CAD data exchange process. Each test procedure was set up with respect to a specific test goal.

VDA IGES SUBSETS

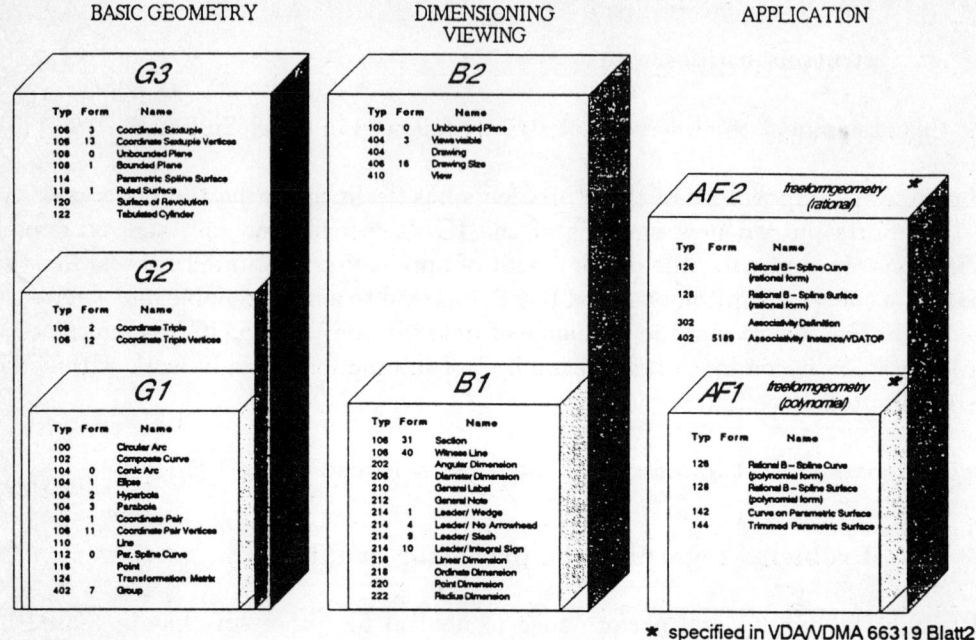

Figure 2. Specification of VDAIS

THE INTER-SYSTEM-TEST

This procedure is the easiest way to get an overview of the expected quality of a CAD data exchange. The neutral file is produced by the pre-processor of system A and afterwards transferred and translated by the post-processor into the receiving system B. In case of errors this test procedure does enable bugs to be traced to either the pre- or post-processor, the tools for neutral file analysis must be used in connection with different CAD models with well known contents (entities, structures, associations, coordinates, presentation attributes) as test data.

THE CYCLE-TEST

The cycle test is normally the first test procedure which is used to get a quick impression of the performance of pre- and post-processor of one system. In this case the IGES data produced by the pre-processor is immediately re-translated

Testmethodology for Neutral File Processors

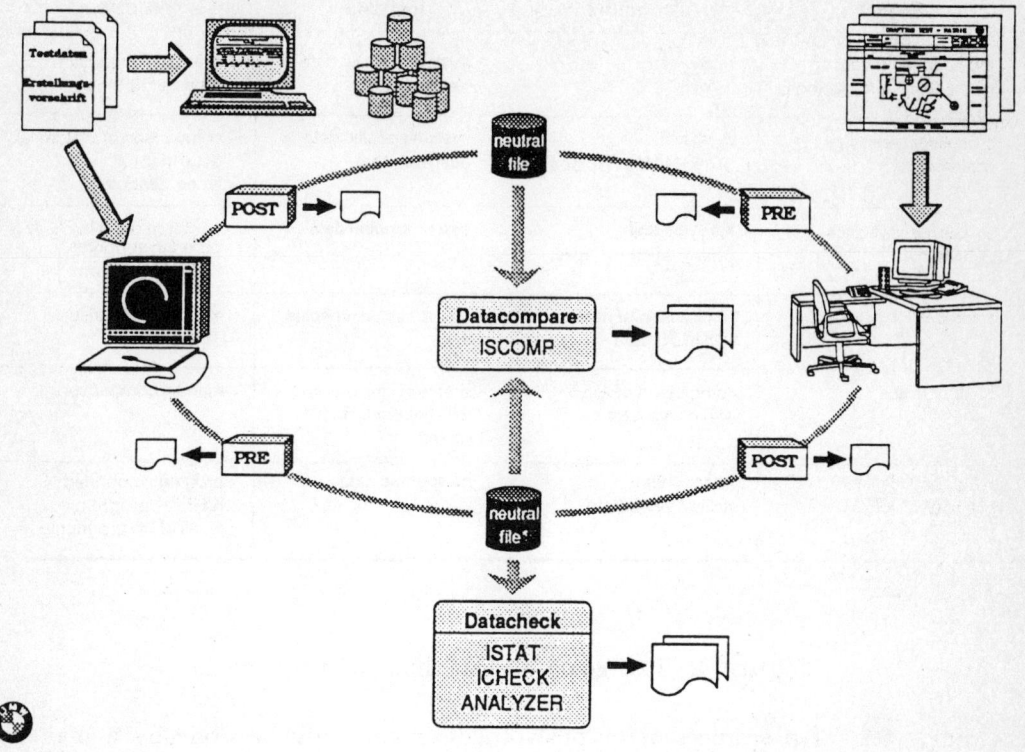

Figure 3. Test tools and test methodology for neutral file processor testing

back into the systems internal format by the post-processor. To trace the source of an error to either the pre- or to the post-processor, a check of the neutral file produced by the pre-processor is required. The test data used is identical to that used within the inter-system-test.

THE FILE-CYCLE-TEST

A special test procedure is the file-cycle-test which requires the existence of 100% correct neutral format test data. In this case the synthetic neutral format test data, which is specified according to a test model generation description, is translated via the post-processor into the system. Using the CAD system functionality the data read in can be compared with the items specified in the generation description. After conversion of the test data back into an IGES format the IGES Statistic Comparator Program ISCOMP provides a quick analysis of the differences between the original IGES file and the file after post- and pre-processing.

IGES PROCESSOR TEST

test goal	test method	test data	evaluation
exchangeable data sets between systems A and B	inter – system – test	system specific data (test matrices)	manual comparison types, attributes, ...
processor capacities of system X	cycle test pre — IGES — post	system specific data (test matrices)	comparison of system data, manual types, attributes
processor capacities of system X	file cycle test post—system—pre	system neutral data	comparison of IGES data program supported ISCOMP program
preprocessor	comparison of native with IGES data	simple system specific data	manual comparison
postprocessor	comparison of IGES with native data	simple system neutral data (synthetic test library)	manual comparison
syntax and statistic check of an IGES file	syntax check statistic evaluation	no specific data	program supported ICHECK program STATISTIC program

Figure 4. Test goals, test methods and test data

A single test of pre-processor or post-processor can only be done by a manual comparison of IGES data and CAD model data. It is obvious that this procedure requires very small test models and should be used only for very special test cases.

Figure 4 summarizes the methodology for neutral file processor testing.

1.4   Problems caused by differences between CAD/CAM systems

Experiences gained with neutral format interfaces showed that in addition to the aforementioned problems, the quality of the data transfer is essentially dependent on the principle differences between the CAD/CAM systems involved in the exchange process. The amount of information lost during an exchange of data is due to differences between the sending and receiving system in:

* CAD model classes implemented (see Figure 5)

  for example: *-2D geometry*

                         - dimensioning, viewing and layout

                         - 3D geometry and ruled surfaces

                         - free-form geometry

                         - solid geometry

* mathematical description of entity types

  for example: *- polynomial form or rational form of curves and surfaces*

                         - implicit definition or explicit definition of conics

* CAD model structures

  for example: *- symbol structure or grouping mechanism*

                         - hierarchy nesting depth

* system and representation accuracy

  for example: *- single precision or double precision*

                         - computing accuracy

                         - value ranges

To overcome the problems caused by system differences a special converter is needed with the ability to adapt neutral files to the capabilities of the receiving system. With the help of such a system, which has been developed within the CAD*I project, an almost complete transfer of the relevant information between CAD systems from different application fields is achievable.

Another advantage of this NEUTRAL FILE ADAPTING SYSTEM (NFAS) is the possibility not only to compensate for system differences but also to manipulate the IGES data with respect to a certain application. This manipulation facility can save a lot of time which in the past had to be spent on adapting the received CAD model to the requirements of the application and to complete the model with information which was lost during the exchange process but which was still needed.

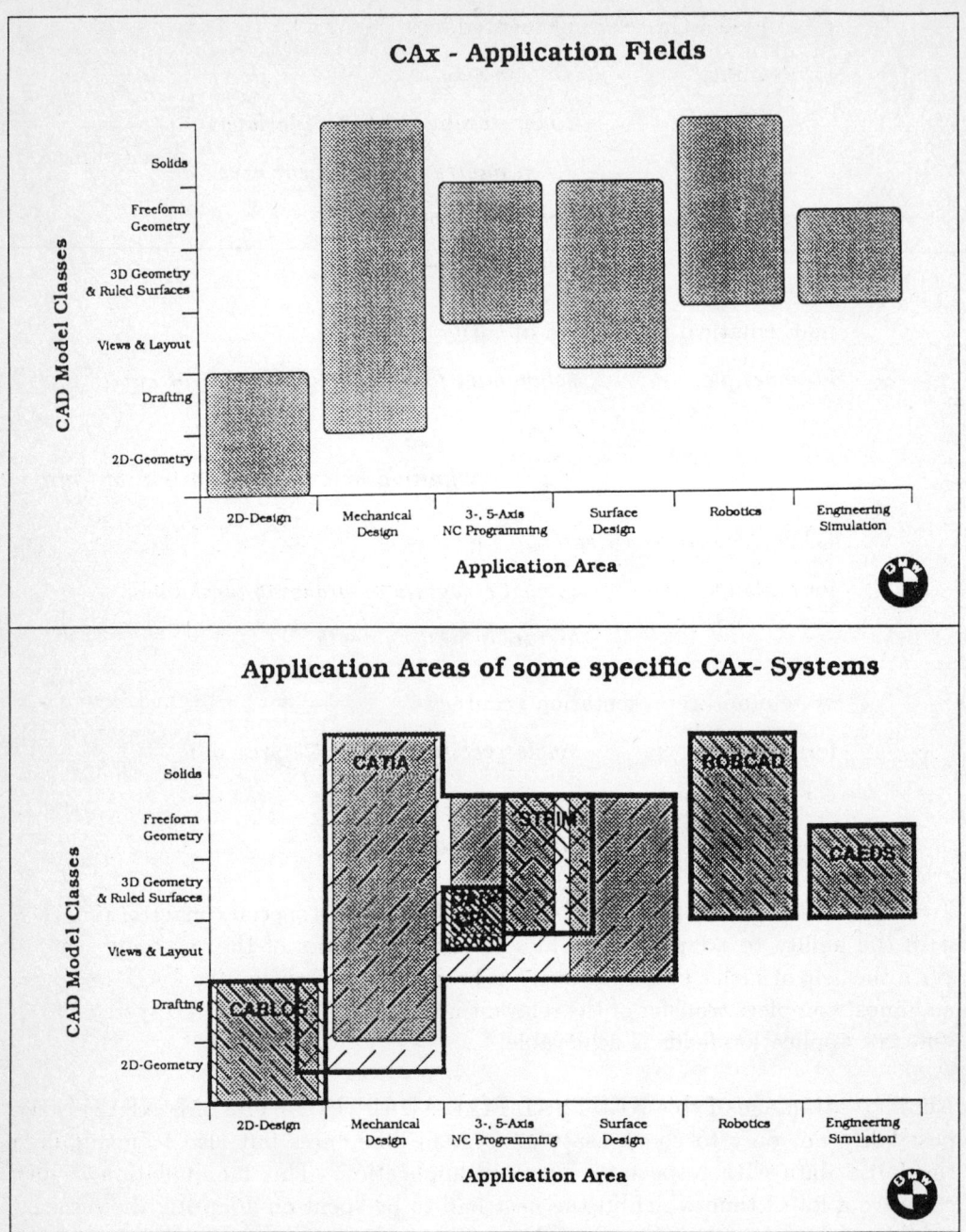

Figure 5. CAD model classes with respect to application areas

1.5   Usage of CAD data exchange software in the industrial environment

To overcome the problems of CAD data exchange, as described above, the development of CAD data exchange software was started in 1984 within the CAD*I project. Since then a large set of tools has been developed and introduced within the data exchange environments of CAD systems in industrial use. Owing that today CAD/CAM data has to be transferred daily, the support of CAD/CAM data exchange is of great importance.

For example, between BMW and its approximately 250 suppliers using more than 60 different CAD/CAM systems, CAD/CAM data has to be exchanged every day. Figure 6 gives an impression of the CAD data flow between BMW and the supplier industries and of the usage of neutral file test tools.

First, after finishing the CAD model to the degree which is required for the moment, a neutral file is produced by the systems pre-processor.

The second step after the conversion is a complete data check of the neutral file in terms of syntactical correctness using the program ICHECK. This means that the whole file is scanned character by character, then assembled into tokens according to the delimiters declared in the global section of the IGES file. Afterwards these tokens and their sequence are checked against the IGES specification. In this way the whole contents of the neutral file is analysed step by step including the structure represented by the large number of pointers necessary in the IGES format.

The contents of a VDAFS file is analysed in an analogous way using the VDAFS test tool ANALYZER.

After this data check the neutral file can be adapted to reach a state whereby optimal information transfer to the receiving system of the supplier is achieved. In Appendix B a detailed introduction to the neutral file adapting is given. To finish the process the data can be archived within CADNET, a CAD model administration system and written onto tape and sent to the supplier.

CAD data sent by suppliers is received at BMW in an analogous way. After being read in, the data is thoroughly checked using the IGES or VDAFS test tools. Following an analysis of the contents of the file and with the knowledge of the sending system, the data is adapted to the capabilities of the receiving system, with regard to the application for which it is needed. Finally, the data is translated via the post-processor into the CAD system.

For further and more detailed information see Appendix A.

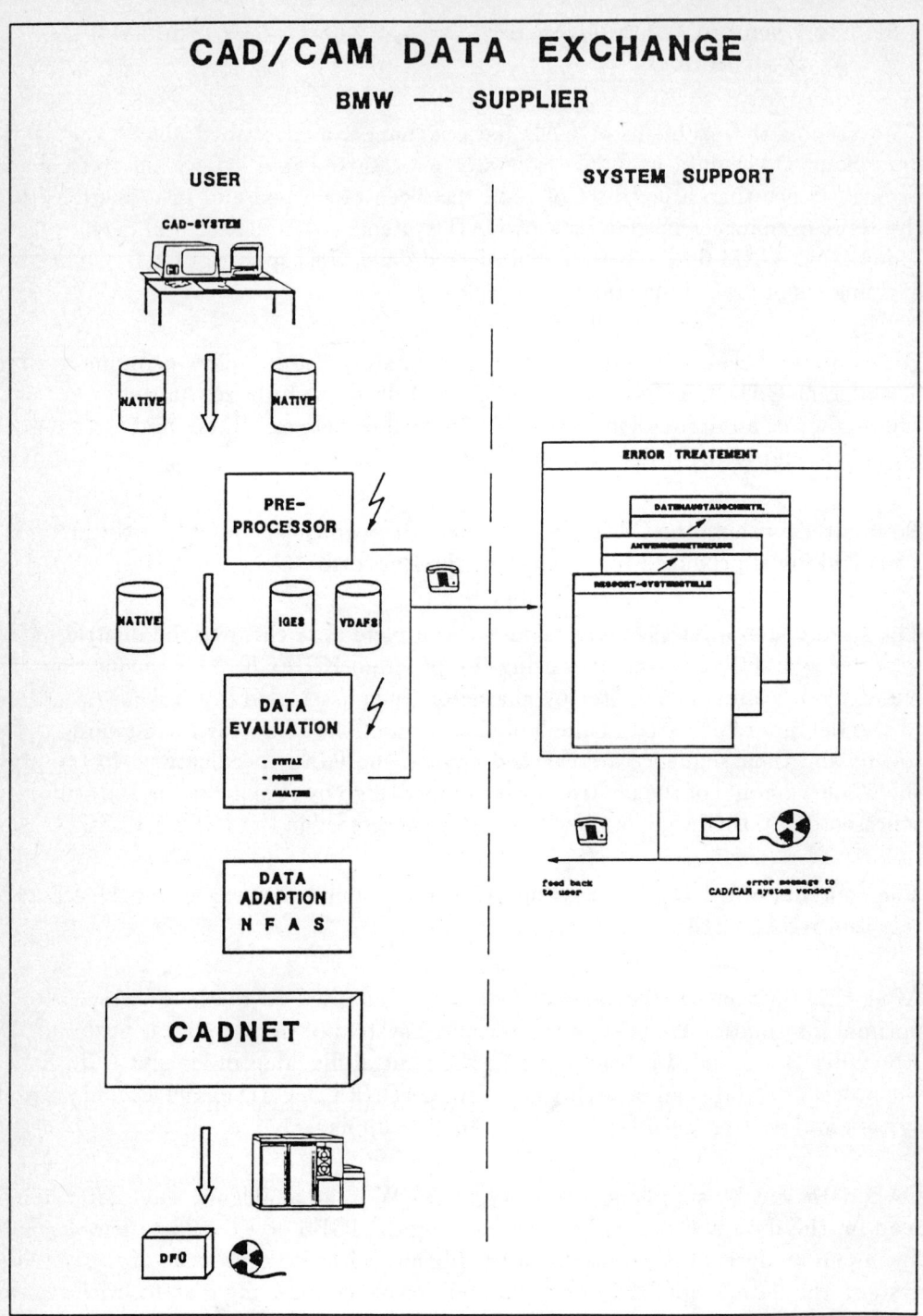

Figure 6. Usage of data exchange software at BMW

1.6 Basis of neutral file adaption

As already mentioned the quality of CAD/CAM data transfer is still restricted because of differences between the applications covered by the systems and also due to principle differences between the various systems.

These principle problems are caused by differences like:

* different geometric modellers
    - 2D / 3D wireframe modeller
    - surface modeller
    - solid modeller

* existence of synonymous entity types definitions in the neutral file specification
    - rational B-spline curve : parametric spline curve
    - rational B-spline surface : parametric spline surface
    - cross hatching : section

* existence of homonymous entity types definitions in the neutral file specification
    - the *copious data* is used as annotation entity as well as geometry entity
    - the *transformation matrix* is used for the definition of geometry entities as well as independent axis system entity.

Therefore two objectives for data adaption have to be taken into consideration.

- First, the data has to be adapted in such a way that it can be handled by the receiving system. By being similar to the data generated by the system itself it can be stored in an identical form within the systems database.
- Second, the data has to be adapted according to the requirements of the application for which it is intended.

These objectives are essential for the quality of the data exchange process.

1.6.1 Overcoming system differences

Normally information (CAD data) is stored within system specific formats and in special descriptions. For example a curve description can be stored in many different ways as it is shown below:

- polynomial forms

- parametric forms

- rational forms

- canonical forms

Even if all representations of an entity type may be exactly transferable into each other, it is clear that a systems processor will not normally be prepared to handle all possible representations of that information, as in the various curve representations described above.

This variety of possible descriptions of one and the same entity will be reduced by using neutral file processors for exchanging data between CAD systems. Nevertheless this fact cannot be fully overcome since there exist redundant entity definitions within the specification itself. Below are listed some occurrences of redundant entities within the IGES specification.

-	Parametric Spline Curve	:	B-Spline Curve
-	Parametric Spline Surface	:	B-Spline Surface
-	Cross Hatching	:	Section
-	Copious Data	:	Group of Points or Lines

This problem cannot be overcome by enlarging the capabilities of the neutral file specification as it will be done by the introduction of STEP (see Figure 7). Even the VDAIS, which can be interpreted as an implementation guide-line for the IGES specification, cannot avoid these adapting problems completely. The solution of the problem described above is the conversion between different representations of the information. This can be done by a system independent adapting program that works on the basis of a neutral format and can therefore be used for any combination of CAD/CAM systems. Figure 8 shows the principle of such a neutral file adapting system.

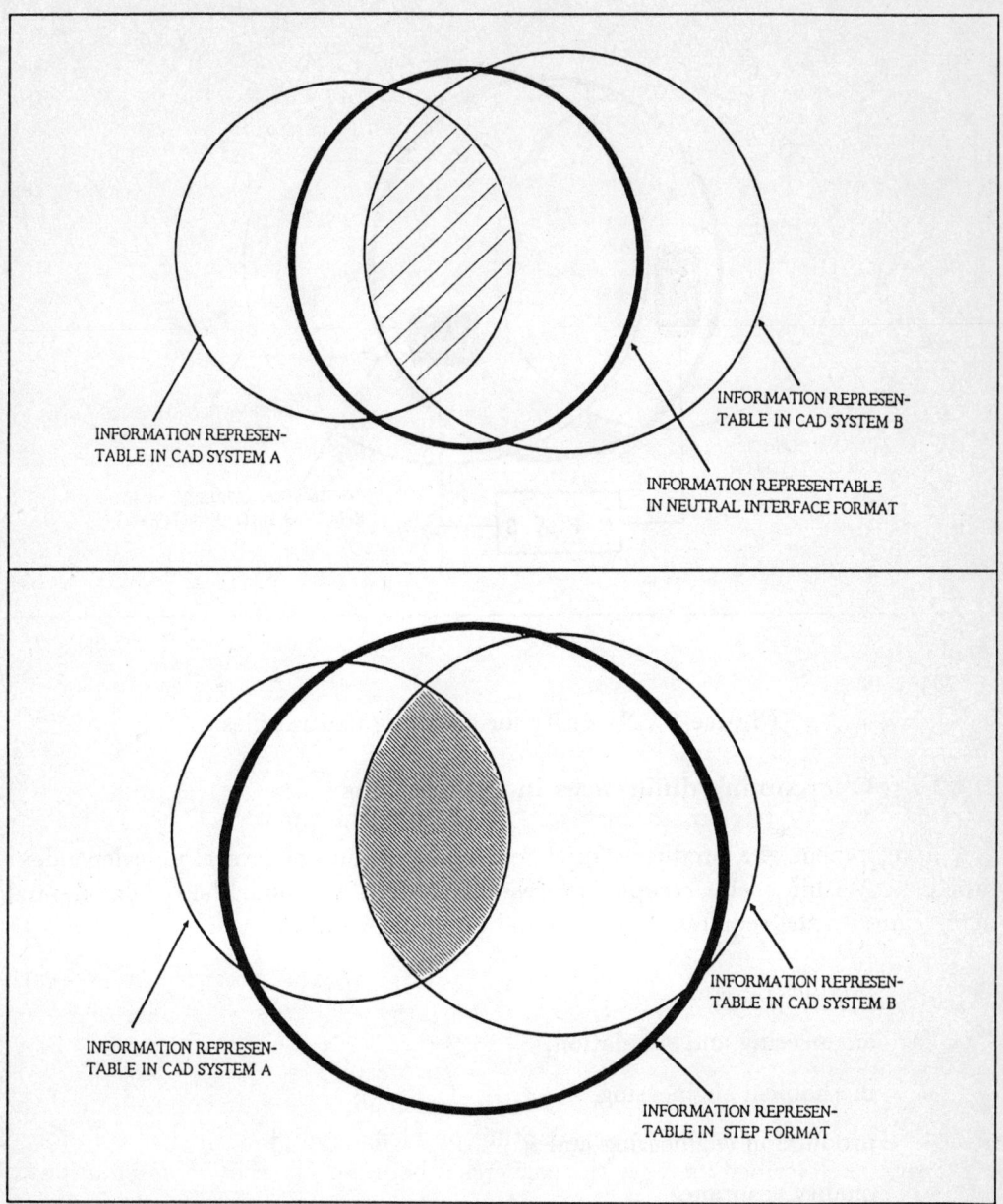

Figure 7. Information transferable via CAD interfaces

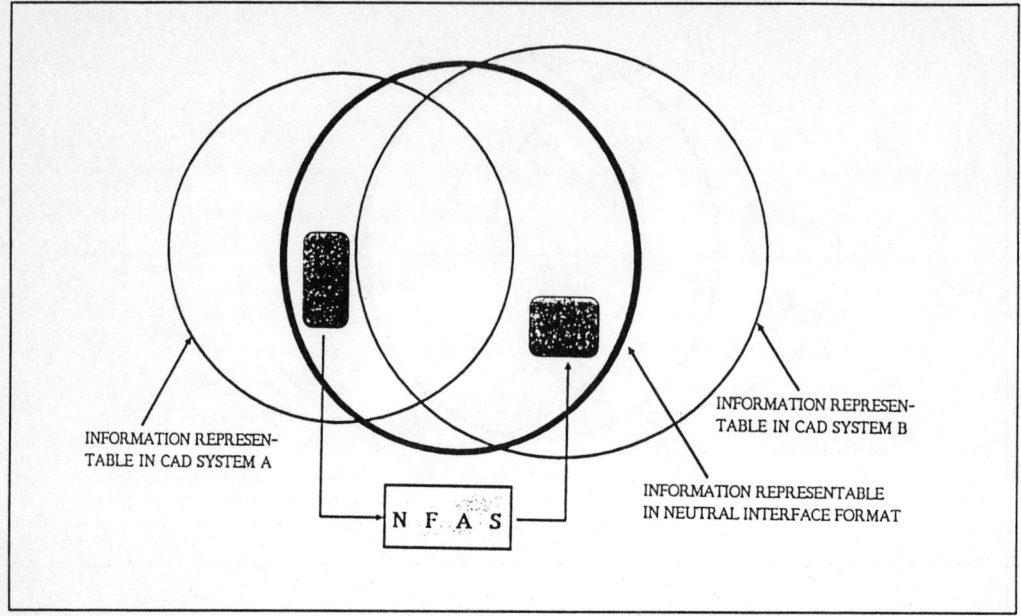

INFORMATION REPRESEN-
TABLE IN CAD SYSTEM A

INFORMATION REPRESEN-
TABLE IN CAD SYSTEM B

INFORMATION REPRESENTABLE
IN NEUTRAL INTERFACE FORMAT

N F A S

Figure 8. Necessity for adapting neutral files

1.6.2   Overcoming differences in applications

The development of a product consists of the connection of several interdependent processes. Within such a complex process chain the CAD model has to be adapted many times to the requirements of special applications like:

- product design,

- engineering and simulation,

- mechanical engineering,

- production engineering and

- quality assurance.

The productivity of CAD systems can be increased enormously by establishing an automatic data adaption. Figure 9 shows the two possibilities of using the Neutral File Adapting System (NFAS). The normal task of NFAS is a mixture of both applications.

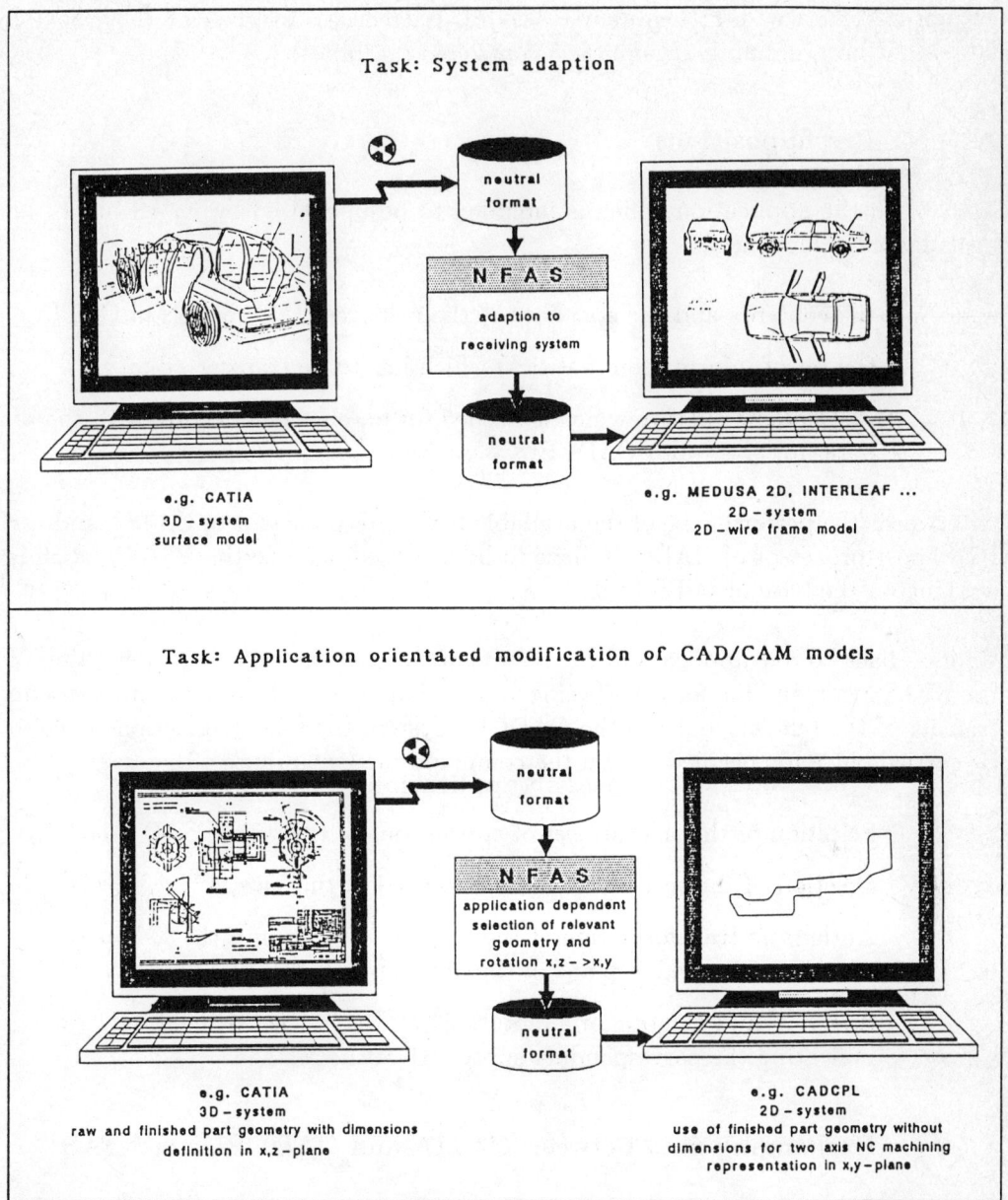

Figure 9. Example for a system and an application adaption by NFAS

1.7 Usage of Neutral File Adapting System NFAS

In order to give an impression of the importance of the neutral file adapting process
we describe the use of NFAS in the transfer of CAD model data from the design

system CATIA to the 2D system CADCPL (the design module of the EXAPT system for programming of numerical controlled machines).

1.7.1 Presuppositions

First of all the application which is intended to be optimised by NFAS has to be investigated regarding

* the contents and the structure of the resource CAD models in CATIA,

* the set of information (entities) which has to be transferred and

* the model structure which is needed for an efficient working with the data imported into CADCPL.

Afterwards the performance of the available IGES pre-processor of CATIA and the IGES post-processor of CADCPL have to be analysed because the NFAS system is working on the base of IGES files.

Coming back to our application, the results of this analysis allow the definition of the NFAS command file for this specific application. In detail, to reach an optimum transfer of the relevant data to the CADCPL system the following actions have to be carried out and specified within the command file (see Figure 10).

- Selection of the relevant set of entities out of the whole CAD model.

- Deletion of the rest of entities not needed in this case.

- Automatic transformation of the selected entities into the X-Y plane, which is the working plane of the 2D system CADCPL.

- Selection and deletion of annotation entities not yet deleted by indicating the corresponding subset B1 of the VDAIS.

1.7.2 Exchanging data between CATIA and CADCPL via NFAS

The original CAD model defined in CATIA which is shown in Figure 11 contains:

- a surface representation,

- a solid representation of the geometry and in addition

- a drawing with a 2D wireframe representation and annotations.

It does not make sense to split up the CAD model into several smaller CAD models

```
!*********************************************************************
! NFAS COMMAND FILE:   C A T I A  -  C A D C P L
!*********************************************************************
!
!  THIS COMMAND FILE IS USED TO ADAPT THE DATA
!  FROM:   CATIA V2R2 PTF4
!  TO:     CADCPL (EXAPT) V 3.8
!
!*********************************************************************
!
! AUTHOR:  H. SCHEDER  FI-100          DATE: 03-06-1989
!
! HISTORY:
!
!*********************************************************************
!
!--------------------------------------------------------------------
! SELECTION OF RELEVENT ENTITY SET AND AUTOMATIC TRANSFORMATION
! INTO X-Y PLANE
!--------------------------------------------------------------------
  PICK LEVEL.EQ.20 PURGE
     AUTOROTATE INTO=XY
  ENDPICK
!
!--------------------------------------------------------------------
! DELETION OF ANNOTATIONS (SUBSET VDAIS B1)
!--------------------------------------------------------------------
  SELECT SUBSET.EQ."VDAIS B1"
     TREEDELETE
  ENDSELECT
!
!--------------------------------------------------------------------
! TERMINATION
!--------------------------------------------------------------------
  EXIT
```

Figure 10. Example for a NFAS command file

containing only the information which can be handled by the systems for applications succeeding the design process.

The original CAD model has now to be transferred to the EXAPT system. After receiving the model it is treated by NFAS according to the command file shown

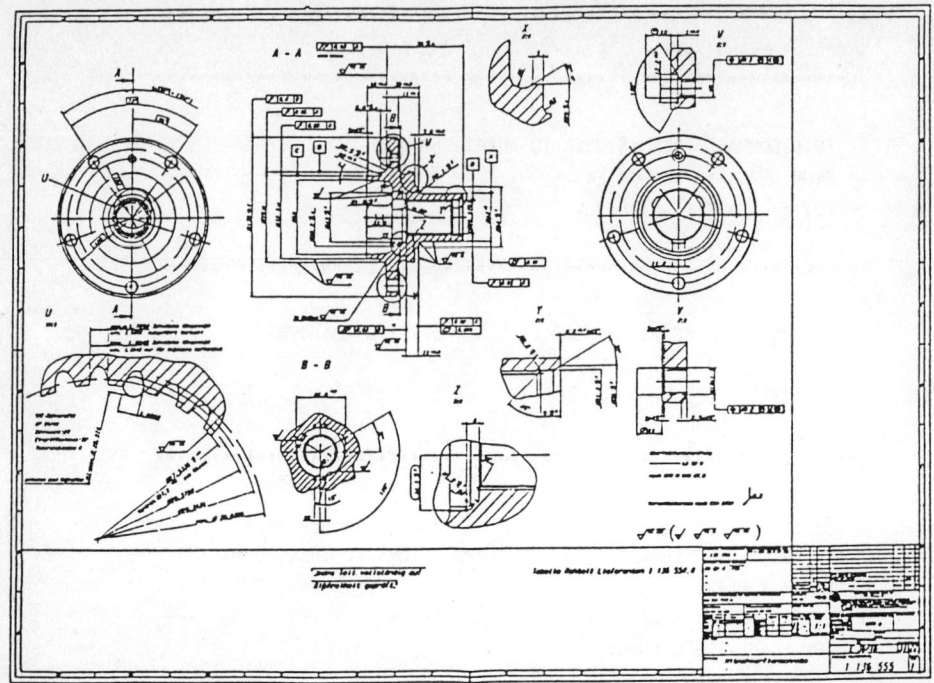

Figure 11. Original CAD Model

in Figure 9 and adapted to the special requirements of the application. Figure 12 shows the result after the translation into the CADCPL system, now ready to proceed with NC programming without subsequent treatment.

1.8   Summary

CAD/CAM data exchange has become important in terms of:

* increasing the productivity of product development,

* shortening the product development cycles and

* improving the product quality.

To support these processes a data exchange infrastructure is strongly required. This means that data produced by processors as well as the neutral file processors themselves have to be validated before they can be used in a productive environment.

These chapters give an introduction to the methodology and software tools needed for an effective support of CAD/CAM data exchange and demonstrates their use within the industrial data exchange processes.

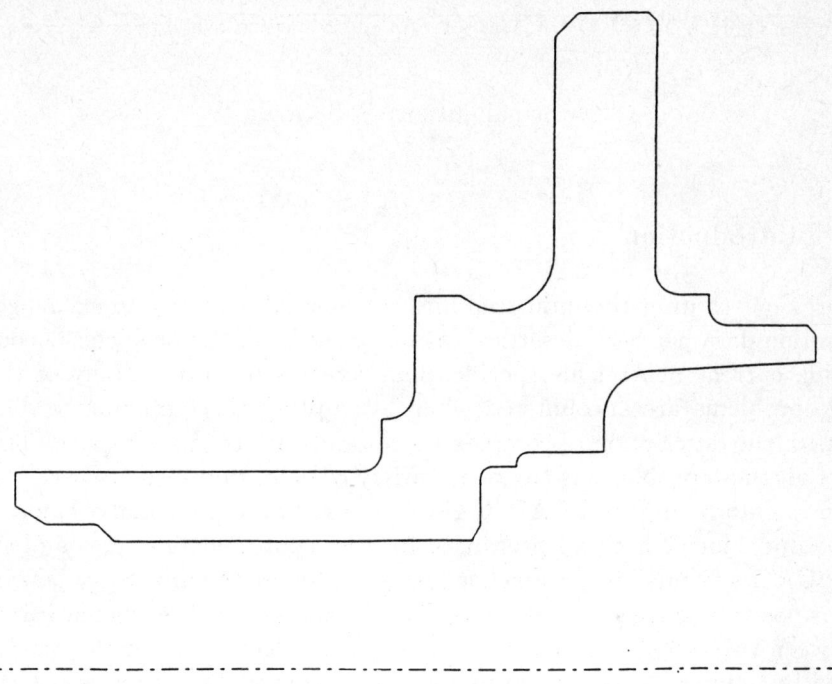

Figure 12. CAD model for NC programming after adaptation by NFAS

As CAD systems are designed for certain applications, problems in exchanging data will remain even if the neutral file processors are free of errors. These principle problems are caused by differences like:

* different geometric modellers,

* existence of synonymous entity type definitions and

* existence of homonymous entity type definitions in the neutral file specification.

A successful approach to overcome these problems described with acceptable efforts was the development of the NEUTRAL FILE ADAPTING SYSTEM NFAS for system and application specific adaption of neutral files.

Principal author: R.J. Goult

2.0   Introduction

In the previous chapter the industrial and commercial necessity to exchange product definition data has been described, as well as some of the problems encountered with using current neutral file specifications for this purpose. Many of the least tractable problems are encountered when attempting to communicate data concerned with the description of complex curves and surfaces in 3D space. The basic source of all these problems is the wide variety of definition formats used for curve and surface data by different CAD/CAM systems. In the past analytic, parametric and procedural forms have all been used for the definitions of curves and surfaces in CAD/CAM systems but parametric forms are by far the most popular. In some cases it is possible to perform, theoretically at least, an exact mathematical conversion between two different forms of representation, algorithms for this purpose are described in Chapter 4. In far too many cases no exact conversion is possible and the best that can be achieved is to produce some form of 'good' approximation to the original data which is acceptable to the receiving system. Some useful approximation algorithms which preserve as much as possible of the geometric intent of the original design are described in Chapters 5 and 6.

2.1   Forms of representation

The basic distinction between analytic, parametric and procedural forms of definition is best illustrated by a simple example. A circle of radius R centred at the origin can be described analytically by the implicit equation

$$x^2 + y^2 = R^2$$

or by the explicit analytic equation

$$y = \sqrt{R^2 - x^2}.$$

The same curve can be described by many parametric equations the simplest of which is

$$x = R\cos t, \quad y = R\sin t,$$

in this case the complete circle is obtained by permitting the parameter t to take all possible values in the range $0 \leq t \leq 2\pi$.

The same curve can be described procedurally as the locus of a point which moves such that the distance from the origin is R.

In this simple example all 3 forms of representation exist and conversion between them is simple, in more complex cases this conversion may be difficult or even impossible. More generally, any equation $f(x,y) = 0$ will represent analytically a curve in 2D Cartesian space, and $\phi(x,y,z) = 0$ will be the general form for the implicit analytic representation of a surface. The analytic form of representation for a curve in 3D space requires two equations $f(x,y,z) = 0$ and $g(x,y,z) = 0$. In effect the curve is described as the intersection of two surfaces and this form of representation is rarely used in CAD systems.

The general form of parametric representation for a curve in 2D space is given by $x = f_1(t)$, $y = f_2(t)$ where f_1 and f_2 are arbitrary functions and t is the parameter. For a 3D curve an additional equation $z = f_3(t)$ is required. For a surface two independent parameters u and v are required and the parametric equations are

$$x = \phi_1(u,v)$$

$$y = \phi_2(u,v)$$

$$z = \phi_3(u,v)$$

Two significant families of curves on the surface are obtained by constraining $u = u_0$ to obtain a parametric curve equation with v as parameter, or by constraining $v = v_0$, to obtain a parametric curve dependent upon u. These families of curves can be used to provide a simple wire-frame visualisation of the surface.

A procedural representation of a curve or surface is any method of describing how to generate points on the curve or surface without giving either an analytic or parametric representation. As an example of such a definition a generalised cone could be described as the surface generated by joining a fixed point (the vertex) to points on a given curve. A less obvious form of procedural definition would be the set of points satisfying a particular differential equation. An important class of curves and surfaces in the CAD context are offsets, an offset curve or surface can be easily described procedurally from a given curve or surface but except in the simplest cases it can rarely be represented as an analytic or parametric curve or surface of the same form as the original.

In modern CAD systems the use of analytic curves and surfaces is usually restricted to the conics (circle, ellipse, parabola, hyperbola) and quadric surfaces (cone, cylinder, sphere, ellipsoid etc) commonly found in solid modelling systems. These particular curves and surfaces are required by a B-Rep modeller which references the curves and surfaces associated with the faces and edges of simple solids. In all these cases simple parametric representations also exist and these are explicitly contained in the CAD*I and STEP neutral file specifications.

Amongst the CAD systems using parametric representations the most commonly used parametric functions are simple polynomials or rational expressions. A simple single segment parametric polynomial curve of degree n is uniquely defined by $n+1$ vector quantities, these may be the simple polynomial coefficients or, of rather more geometric significance the control points for the corresponding Bézier or B-Spline curve. More generally a composite piecewise polynomial curve can be represented by the polynomial coefficients of its individual segments, as a piecewise Bézier curve specifying the control points for each segment or as a single B-Spline curve specified by the control points and the knot set. Three equivalent forms of representation, explicit coefficient, Bézier or B-Spline, exist for parametric polynomial surfaces. By using homogeneous coordinates including a fourth 'weight' coordinate the three forms of representation described above can be generalised to define rational curves and surfaces. With this convention each point P in space is associated with the coordinates (x, y, z, w) but only the ratio of these quantities is significant; the corresponding Cartesian coordinates (X, Y, Z) are $X = x/w, Y = y/w, Z = z/w$. If each of x, y, z, w is then defined as a parametric polynomial function of a particular degree then the resulting curve is a rational curve of that degree. It should be noted that a polynomial curve or surface is a special case of the rational form in which w has the constant value 1.

For these commonly used forms of parametric curve and surface representation exact mathematical conversion is possible only if the systems are either both rational or both polynomial and have the same maximum degree. In these particular cases Chapter 4 gives details of algorithms for conversion between the Bézier, B-Spline and explicit coefficient forms of representation. In all other cases data exchange is only possible with some form of approximation. For the transfer of data from a system using a high degree polynomial representation to a system with a lower limit on the permissible degree this approximation takes the form of degree reduction which is described in Chapter 5. Chapter 6 describes algorithms suitable for other more general types of approximation problem including rational to polynomial conversion and the communication of some types of procedural curve and surface data. Two important features of the algorithms described in these chapters are the ability to obtain a bound for the approximation errors and the ability to maintain continuity conditions at the ends of curve segments or along the edges of surface patches. These two properties make it possible to develop procedures which will automatically subdivide a curve or surface in order to produce a piecewise approximation which maintains the required level of continuity and produces errors less than a specified tolerance.

2.2   Exchange mechanisms

Three different strategies have been widely used for the exchange of geometric data between CAE systems. These strategies are to develop dedicated direct system to system translators, to make use of some agreed neutral file format for intermediate data communication or, as a totally different approach to utilise a parametric evaluator.

Direct system to system translators can in theory ensure the complete communication of all compatible data between the two systems concerned. For any pair of systems two translators are required, one in each direction, and these will both require modification if either system changes. The disadvantages of this approach become more apparent as the number of communicating systems increases. For N systems N(N-1) direct translators are required and 2(N-1) of these will require modification if any system changes. This situation is illustrated in Figure 13.

As the number of supported systems increases the neutral file approach offers the advantage of only requiring 2 translators for each system supported hence for N systems 2N translators are required. If one system is modified only the 2 translators associated with it will require modification. The translators associated with a particular system are a pre-processor which converts internal system data to the neutral file format and a post-processor which accepts data from the neutral file and converts to the system's internal format. Figure 14 shows the translator requirements for a number of communicating systems. The disadvantage of using a neutral file is that only data represented in both systems and in the neutral file format specification can potentially be communicated exactly. In practice there may be further restrictions imposed by the capabilities of particular pre- and post-processors.

For either of the above approaches data conversion will be required when the communicating systems contain mathematically equivalent representations of the same geometric data. Where the geometric representations are incompatible (eg communication between rational and polynomial based systems) the use of some form of approximation is unavoidable. With the neutral file approach another consideration is the geometric representation required by the neutral file, this may require conversion or approximation to occur in the pre- or post-processor even when the communicating systems have identical and internal geometric representations. Chapter 3 considers the requirements for an ideal neutral file specification in rather more detail.

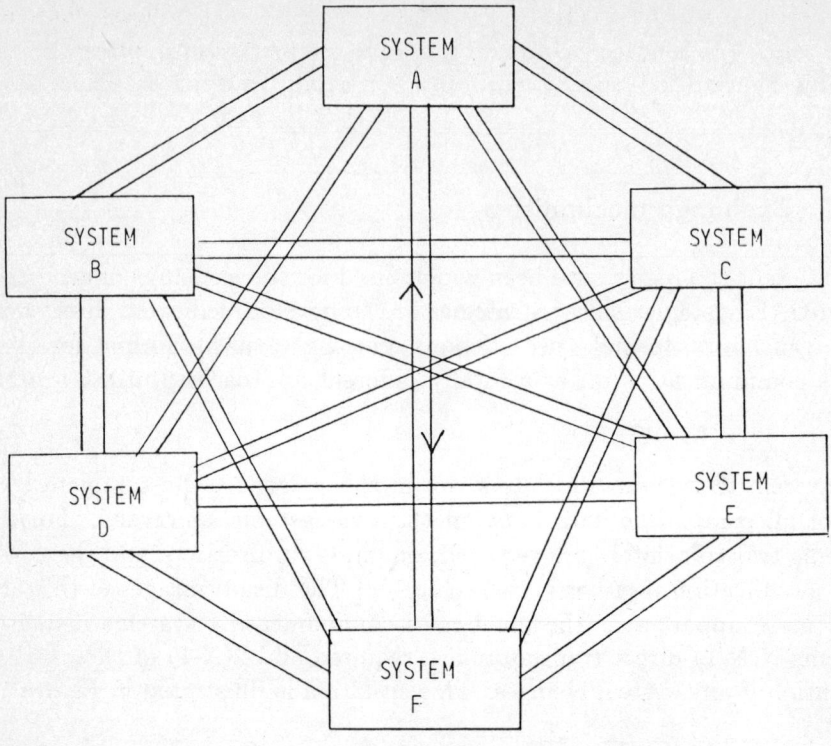

Figure 13. Direct translator requirements

2.2.1 Parametric Evaluators

Parametric evaluators provide a form of applications interface to CAD surface design systems. The concept differs significantly from the neutral file approach in that a complete and explicit surface definition is never communicated. For each type of surface communicated a 'black box' of software is provided to perform the surface evaluation function, this is accompanied by data to define the parameters of the particular surface being evaluated. All surfaces must be parametric of the form $\mathbf{r}(u,v)$ and the evaluator is a function or procedure which for the input of parameters u and v returns the values of the point $\mathbf{r}(u,v)$ and, usually, the first and second partial derivatives at this point. The data returned is sufficient for many applications including surface display, for which point and surface normal are required, NC machining computations, surface offset calculations and the computa-

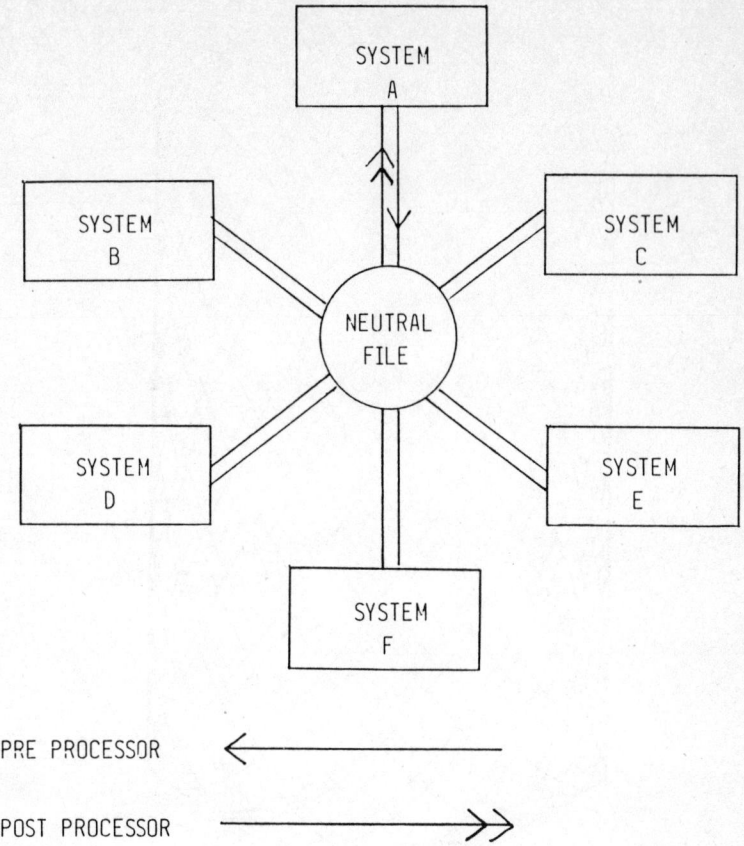

Figure 14. Translator requirements for neutral file interfaces

tion of geodesics on the surface. One significant advantage of parametric evaluators is that the surface data is always in theory communicated exactly with none of the errors associated with the type of approximation process described in chapters 5 and 6.

The major disadvantage of parametric evaluators is that the application software must be specially written to make use of the data in this form. Successful parametric evaluator interfaces have been developed at Cranfield to the APT IV Sculptured Surfaces system for NC machining and to the companion CASPA graphics display system. A parametric evaluator based package to compute geodesic paths on a surface for filament winding applications has also been developed (Figure 15). In the United States Northrop and General Dynamics have also used the parametric evaluator concept for joint projects involving geometric definitions from incompatible surface design systems. For their applications the confidentiality of the geometric definition when communicated in this way is considered to be a major advantage,

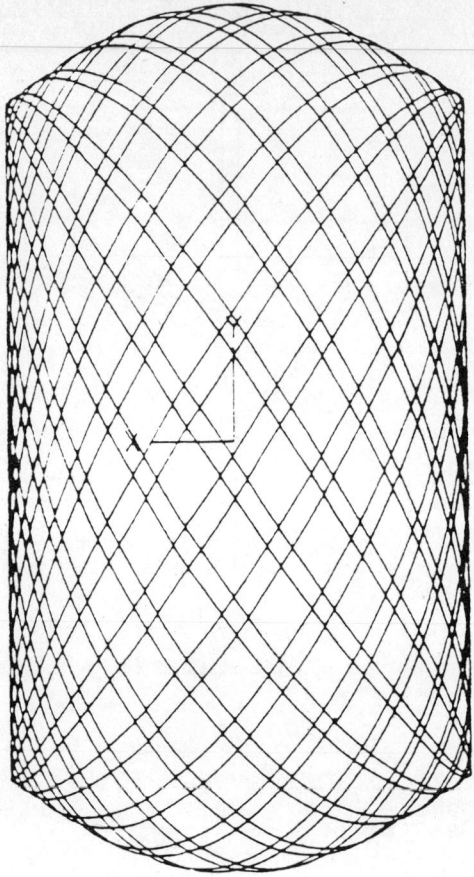

Figure 15. Geodesic curves on a surface computed using a parametric evaluator

as is the inability of the receiving system to modify the geometry. A further disadvantage of parametric evaluators as an exchange mechanism is that although the communication is very precise the receiving system can effectively only examine one point of the surface at a time and receives no explicit indication of overall size and shape. The accompanying data is only meaningful to the generating system and the receiving system cannot modify the geometric definition in any meaningful way.

The inclusion of parametric evaluators in a later version of STEP is being proposed but there are some significant problems to be overcome. In order to standardise the concept an agreed standard format for curve and surface data to be used by the evaluators has to be devised and the not insignificant problems of communicating evaluators between different computers and operating system have to be solved.æ

3 NEUTRAL FILE INTERFACE REQUIREMENTS

Principal authors: R.J. Goult, M.A. Lachance

3.0 Introduction

Ideally a neutral file interface for CAD data exchange should make it possible to communicate 100% of the data between any two systems without loss of accuracy or information. The effectiveness of this transfer depends not only upon the neutral file specification but also upon the quality of the pre- and post-processors provided. In the context of sculptured surface data fundamental incompatibilities between the forms of surface representation used by different systems mean that no matter how good the neutral file specification and the processors the objective of 100% interchange between all systems will never be attainable. The practical limit to the amount of data which can be fully exchanged is the intersection of the capabilities of the two systems, the neutral file specification and of the processors concerned. The design of the neutral file is important in that it defines the ultimate scope of the exchange and strongly influences the quality of pre- and post-processors. This chapter will describe some of the important considerations in defining a neutral file and describe how these have influenced the design of the CAD*I neutral file and the ISO draft proposal STEP. The emphasis is on those entities in the neutral file concerned with the communication of parametric curve and surface data.

3.1 Design objectives

The major objectives in neutral file design can be loosely defined as breadth of scope, efficiency, stability of representation and the provision of unambiguous definitions. Of these objectives efficiency is both the hardest to define and the one which introduces conflicting requirements which ensure that any neutral file design is always a compromise.

A very narrow interpretation of efficiency is that the neutral file communicating any given set of product data should be as small as possible. The problem with this as a sole objective is that a very compressed physical file format is likely to complicate the development of pre- and post-processors and make human interpretation of the contents of the data file virtually impossible. The IGES specification includes provision for compact binary files but this feature is not supported by the majority of commercial pre- and post-processors. A broader interpretation of efficiency should take into account not only file size but also the ease with which pre- and post-processors can convert data between the CAD systems and the neutral file. One pass processing can be an objective and to enable this an important feature of the CAD*I neutral file specification is the avoidance of forward pointers. The breadth of scope of the entity set also influences the ease of processor development and the overall effectiveness of the exchange process. A large entity set with multiple representations of equivalent data makes it possible to match closely the internal representations of a variety of CAD systems and simplifies pre-processor development since direct mappings are provided. This approach, although ideal for archiving, will not however improve the overall efficiency of the exchange process since post-processor development is then more complex and there is a tendency for different systems to support only those entities close to their internal representations. The opposite approach is to insist that there is no redundancy in the physical file representation, but this in turn can produce complex representations of basically simple data occupying an unecessarily large amount of file space. The advantage of a minimal entity set is that different systems are forced to use a common neutral file representation for equivalent data and there is less danger of producing representations not recognised by particular post-processors. In both the CAD*I neutral file and in STEP the entity set is nearly minimal but there are a few exceptions which offer particularly compact representations. One example of this type of compromise is the polyline, simply represented by a sequence of points but essentially redundant since the same information could, with more complexity, be communicated as a B-Spline curve of degree 1, or as a composite curve.

Early investigations in the CAD*I project of IGES pre- and post-processors available with CAD systems showed that at least some of the communication problems were due to the different processor developers using a different interpretation of the entity definitions. This problem is always potentially present when a natural language like English is used for the standard specification. For the CAD*I neutral file specification the formal computer sensible language HDSL (Higher Data Specification Language) was used to produce precise and unambiguous definitions. For STEP a similar computer sensible language EXPRESS has been developed and is included as part of the standard itself. This approach cannot of itself ensure that the definitions themselves are correct or appropriate but they are only capable of one interpretation and the syntax can be verified by suitable software tools. Examples of HDSL and EXPRESS definitions for simple geometric entities are included in the next section.

3.2   Neutral file entities

In defining the individual entities to represent the shape and size of a product in a neutral file a primary consideration is the stability of the representation. Most geometric entities are represented in the neutral file essentially by a sequence of real numbers. During the communication which includes the pre-processor, the neutral file and the post-processor these real numbers may be rounded or truncated and hence the data communicated is rarely precise. A stable neutral file representation for a particular geometric entity should be robust in the presence of such small errors. Two particular aspects of this stability are:

(i)   the post-processor should be able to unambiguously interpret the received data as being of the same geometric form as the original entity.

(ii) for small transmission errors the geometry reconstructed by the receiving system should be close to that originally defined.

Not all entities in existing standards satisfy the above requirements for stability. As an example IGES [14] defines a circular arc (Entity Type 100) in the xy plane by its centre point together with start point and end point for the arc. It is not clear how this data is to be interpreted if the received data is such that the distance from the start point to the centre is not identical to the distance from the centre to the end point.

In selecting the geometric entity representations for STEP and for the CAD*I neutral file stability was a primary consideration. Both STEP and the CAD*I neutral file specification contain entities to represent conic sections and simple surfaces such as sphere, cone, cylinder and torus, in addition to more general parametric curve and surface entities. Both neutral files separate the position and orientation information from the essential geometric properties of the curve or surface being represented. For stability most of the data is in the form of points and directions. Similarities and differences between the two representations and their formal description languages are best illustrated by a simple example.

In the CAD*I neutral file specification [20] an ellipse is described as:

```
ENTITY ELLIPSE=GENERIC (d:DIM)
    STRUCTURE
            semi_major    :      ANY(REAL):  (* positive)
            semi_minor    :      ANY(REAL):  (* positive)
            centre        :      ANY(POINT(d));
            CASE d OF
                D3:  (normal: ANY(DIRECTION(D3)));
                D2:  (normal: NIL);
            END;
            reference_point :    ANY(POINT(d));
            END;
```

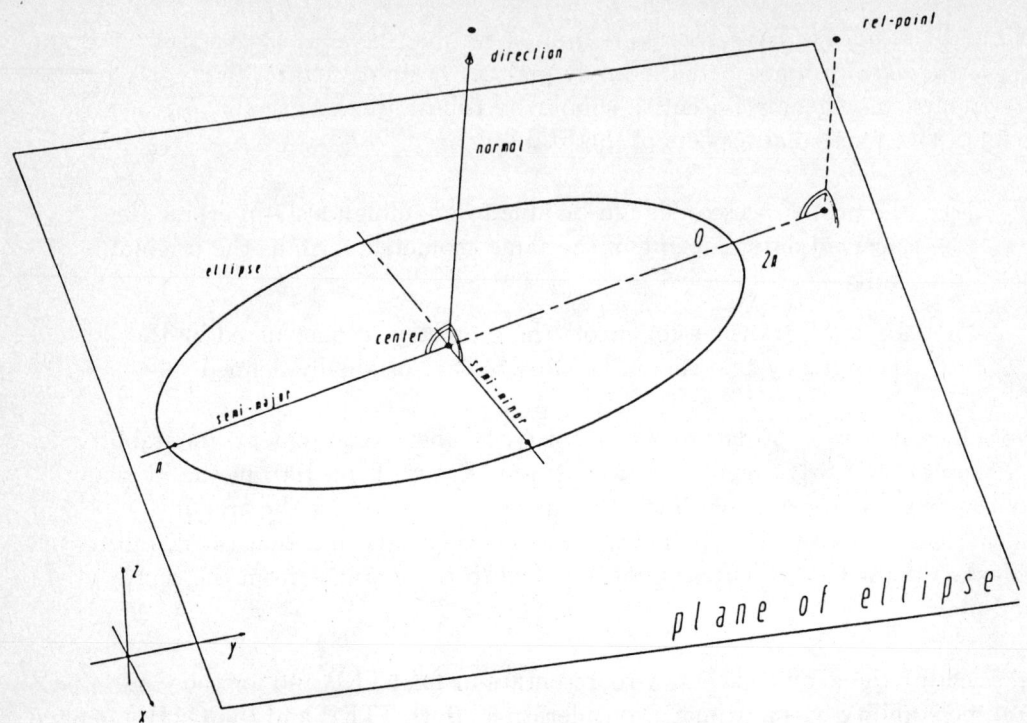

Figure 16. Interpretation of CAD*I HDSL attributes of Ellipse entity

From this definition in order to obtain the parametrisation in the 3 dimensional (D3) case define:

dir=reference_point−centre−((reference_point−centre).normal) normal

Normalise **dir** to produce **a**, a unit vector in the plane of the ellipse, then the parametric equation is

$$\mathbf{r}(u) = \mathbf{centre} + \text{semi\_major } \mathbf{a}\, cos(u) + \text{semi\_minor } (\mathbf{n} * \mathbf{a})\, sin(u).$$

In this equation **n** is a unit vector in the direction of the normal. Figure 16 illustrates the interpretation of the attributes of this entity. In the STEP specification [21] the corresponding entity definition is given by:

> ENTITY ellipse
>
> SUBTYPE OF (conic);
>
> semi_axis_1   :   REAL;
>
> semi_axis_2   :   REAL;
>
> position   :   axis_2_placement;

WHERE

        semi_axis_1>0.0;

        semi_axis_2>0.0;

        coordinate_space(ellipse)=coordinate_space(location);

END_ENTITY.

Within this definition the axis_2 placement entity is the entity used to locate a point and a pair of mutually perpendicular axes in space. It is defined in terms of a point and two directions representing the axis (in this case normal to the plane of the ellipse) and a reference direction. The STEP definition for a 3D entity projects the reference direction onto the plane normal to the axis $\mathbf{a}$ to obtain a unit vector $\mathbf{x}$, then defines $\mathbf{y} = \mathbf{a} * \mathbf{x}$. For a 2D entity $\mathbf{x}$ is obtained by normalising the reference direction and $\mathbf{y}$ is perpendicular to $\mathbf{x}$. If $\mathbf{c}$ is the point defined in the axis_2 placement the parametric equation of the ellipse is defined in the STEP specification as:

$$\mathbf{r}(u) = \mathbf{c} + \text{semi\_axis\_1}\, cos(u)\, \mathbf{x} + \text{semi\_axis\_2}\, sin(u)\, \mathbf{y}$$

It should be noted that both these entities provide a common format for the definition of an ellipse in either 2 dimensional space (D2 in CAD*I, coordinate_space = 2 in STEP) or 3 dimensional space and the definitions define an unambiguous interpretation for all legal values of the data, even if rounding errors produce a reference point or reference direction not strictly in the plane of the curve.

The other simple curve and surface entities in both specifications are defined in a similar way. One example of an entity where the definition has been modified from the usual one in order to improve the numerical stability is the conical surface. The geometric properties of such a surface would appear to be simply defined in terms of a vertex position, an axis and a semi-angle. In practice the entire infinite conical surface is rarely required and the region of interest may be some part of the surface at a considerable distance from the vertex of the cone. In such cases the effective radius and position of the surface section are very sensitive to the values of the semi-angle and the axis direction. For this reason the selected attributes in the CAD*I neutral file are a point, the axis direction, the semi-angle, the radius and a reference point. The point lies on the axis of the cone and the radius gives the radius of a frustrum of the cone at this point. The reference point is used to determine a direction from the axis point which is perpendicular to the axis and is used in the parametric representation of the surface. The corresponding STEP entity uses the radius from a particular point on the axis in a similar way. Clearly these definitions correspond to the elementary one if the axis point is the vertex with corresponding radius of 0.0 but for a surface section some distance from the vertex a more stable representation is available if the axis point is selected close to this section.

3.3 General parametric curve and surface representations

Amongst CAD systems there is a wide variety of formats used to represent general parametric curves and surfaces. For parametric polynomial curves and surfaces these include power series coefficients, B-Spline, Bézier and Ferguson or Hermite. For a single segment curve of a given degree these representations are mathematically equivalent but the data content differs considerably. In the power series form the explicit polynomial coefficients are stored, in the B-Spline and Bézier representation the curve is defined in terms of a set of control points and the appropriate basis functions whilst the Ferguson representation utilises end points and derivatives together with Hermite basis functions. For rational curves and surfaces the power series, B-spline and Bézier formats can be generalised by introducing homogeneous coordinates with a fourth 'weight' coordinate.

Existing standards have different levels of support for these representations. In VDA-FS [23] power basis coefficients are used to express polynomial data. The same entities in IGES are communicated sometimes using polynomial coefficients (entities 112, 114), and sometimes using Rational B-spline coefficients (entities 126, 128). SET [22] uses power basis coefficients and permits a rational form of this representation.

Which representation is 'best' for the set of parametric polynomial curves and surfaces is an issue which has yet to be resolved. Literature which supports one representation over another has been slow to appear. In order to determine the stability of these alternate forms of representation some experiments were organised during the CAD*I project. These experiments are described in the following paragraphs.

3.3.1 Stability experiments

We assume that both the sending and receiving systems employ explicit power basis coefficient representations of polynomials. We assume that there is some loss of precision due to the use of an ASCII format. For the purpose of these experiments, the errors were of three types: truncation to 6 significant figures, truncation to 12 significant figures, and truncation to 4 decimal place accuracy. It is further assumed that all local calculations are performed in double precision.

The polynomials $\mathbf{P}(u)$ to be transmitted could have been arbitrarily chosen. To give them some physical significance, they were chosen to be least squares approximates, of different degrees, to five separate curves in the xy-plane. Three of these were

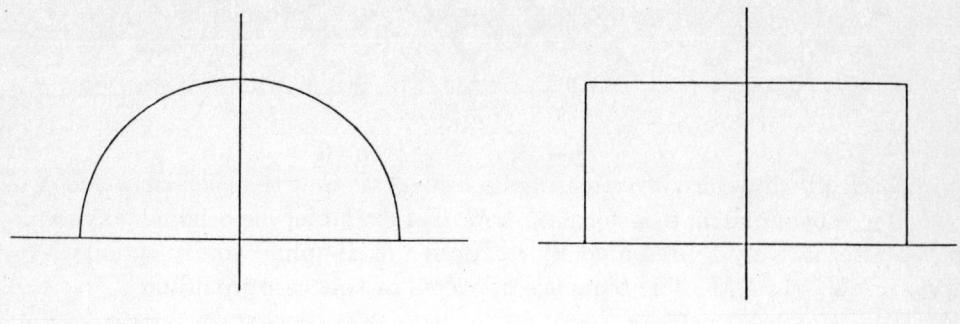

a) $\{\ (x,y)\ :\ x^2 + y^2 = 1,\ y > 0\ \}$ **b)** $\{\ (x,y)\ :\ max(x,y) = 1,\ y > 0\ \}$

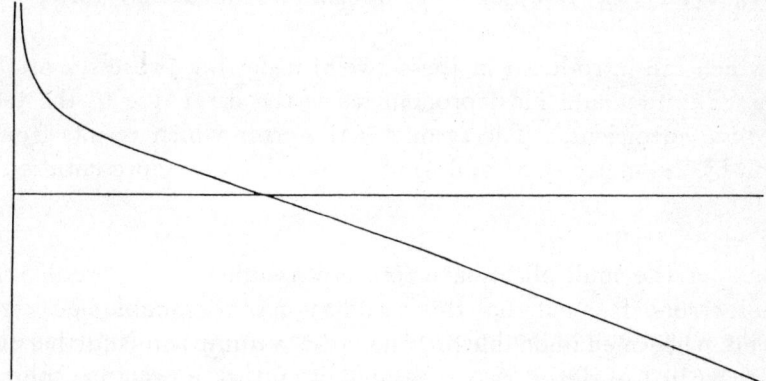

c) $\{\ (x,y)\ :\ x = e^{5u},\ y = 1 - u(2 - u(3 - u(4 - u))),\ 0 \le u \le 1\ \}$

Figure 17. Curves defining original polynomials

semi- circular arcs with different centers and radii, one was the upper half of the unit square, and the last was a mixed exponential and polynomial curve. Some of these curves are displayed in Figure 17.

Figure 18 shows the various communication paths involved. The first path assumes that $\mathbf{P}(u)$ is communicated directly using explicit coefficients, in much the same way as the German automotive standard VDA-FS works today. However, the target system does not receive the polynomial $\mathbf{P}(u)$, but the truncated polynomial $\mathbf{TP}(u)$.

The second path assumes that the Bézier control points will be the medium of exthen the target system will receive this coefficient vector with an associated error vector $\mathbf{a} = (\mathbf{a}_0, \mathbf{a}_1, \ldots, \mathbf{a}_n)$,

$$\mathbf{A} = \mathbf{A} + \mathbf{a}.$$

The approach which we are investigating here suggests that the coefficient vector $\mathbf{A}$ be first altered by a matrix transformation M. For example, the original polynomial might be alternatively represented by its Bézier or B-spline control points $\mathbf{V} = (\mathbf{V}_0, \mathbf{V}_1, \ldots, \mathbf{V}_n) = \mathbf{A}M$. The sequence of events in this case would be

$$\mathbf{A} \longrightarrow \mathbf{V} \longrightarrow \mathbf{V} + \mathbf{v} \longrightarrow \mathbf{A} + \mathbf{v}M^{-1},$$

where $\mathbf{v} = (\mathbf{v}_0, \mathbf{v}_1, \ldots, \mathbf{v}_n)$ represents the associated truncation error.

The errors which are introduced in these two transferring processes are basically of two types. The first and most pronounced is the error due to the truncation of the respective coefficients. The second is the error which results from matrix multiplication by the matrices M and M^{-1}, converting one representation into the another.

In this discussion the multiplication errors are assumed to be small relative to the truncation errors. It is felt that this assumption is reasonable because all local calculations are performed in double precision; this assumption is further supported by the fact that, in the Bézier case, the matrix and its inverse can be expressed explicitly, allowing the calculations to be performed optimally. Even in the B-spline case, where one numerical inversion must take place, the transformation matrices seem to be reasonably well-conditioned when the degree is small [5].

For these reasons we focus on the truncation vectors $\mathbf{a}$ and $\mathbf{v}M^{-1}$. For a polynomial $\mathbf{a}(u)$ parametrised over the interval $[0,1]$, with coefficient vector $\mathbf{a}$, it is easy to show that a uniform bound on $\mathbf{a}(u)$ is

$$\|\mathbf{a}(u)\| \leq (n+1) \max_{0 \leq i \leq n} \|\mathbf{a}_i\| \quad \text{for} \quad 0 \leq u \leq 1,$$

where $\| \; \|$ denotes the Euclidian norm in three space.

Since the Bernstein polynomials and the B-spline basis functions form a partition of unity, it is also easy to see that for a polynomial $\mathbf{v}(u)$ with control polygon $\mathbf{v}$ we have a uniform bound

$$\|\mathbf{v}(u)\| \leq \max_{0 \leq i \leq n} \|\mathbf{v}_i\| \quad \text{for} \quad 0 \leq u \leq 1.$$

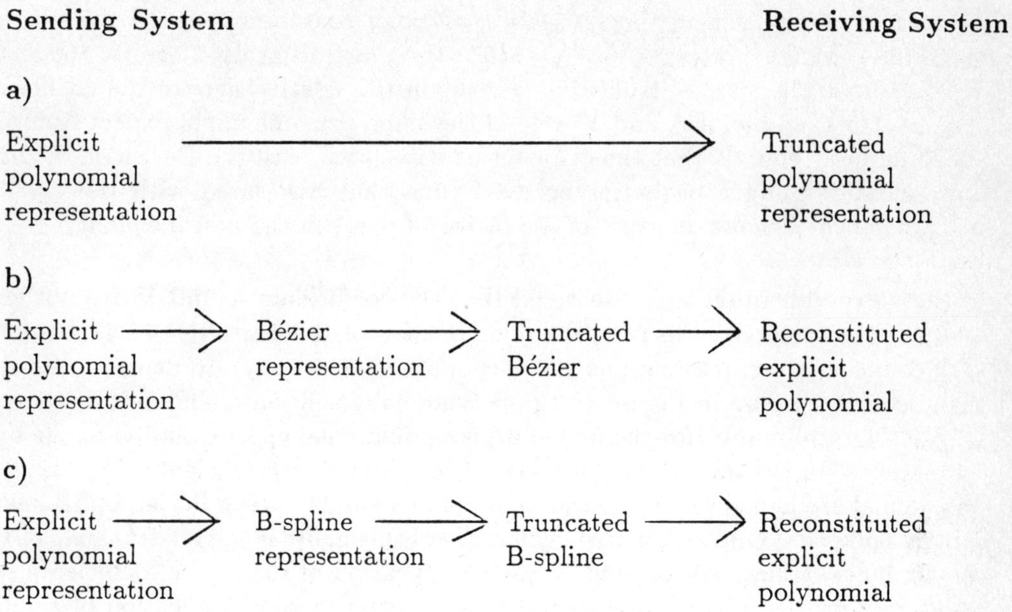

Sending System   Receiving System

a)

Explicit ————————————————————→ Truncated
polynomial   polynomial
representation   representation

b)

Explicit ——→ Bézier ——→ Truncated ——→ Reconstituted
polynomial   representation   Bézier   explicit
representation   polynomial

c)

Explicit ——→ B-spline ——→ Truncated ——→ Reconstituted
polynomial   representation   B-spline   explicit
representation   polynomial

Figure 18.   a) Polynomial coefficients as exchange medium   b) Bézier control points as exchange medium   c) B-spline control points as exchange medium

change. Thus the polynomial $\mathbf{P}(u)$ is transformed to its Bézier representation $\mathbf{B}(u)$, these control points are then truncated to yield $\mathbf{TB}(u)$, and then the receiving system reconstitutes these truncated Bézier control points into an explicit polynomial $\mathbf{RTB}(u)$.

The third path is similar to the second, except that uniform B-spline control points are employed. In this case the polynomial $\mathbf{P}(u)$ is transformed to its B-spline representation $\mathbf{S}(u)$, these control points are truncated to get $\mathbf{TS}(u)$, and then the truncated B-spline control points are reconstituted to form the explicit polynomial $\mathbf{RTS}(u)$. The splines used here are the normalised uniform B-splines $N_{i,n+1}(u)$ of order $n+1$, with distinct integer knots $(i - n - 1, \ldots, i)$, where $i = 1, \ldots, n+1$.

3.3.2   Analysis

It seems quite natural to exchange polynomial data by exchanging the actual polynomial coefficients. The only error that is incurred is a truncation error. That is, if $\mathbf{A} = (\mathbf{A}_0, \mathbf{A}_1, \ldots, \mathbf{A}_n)$ represents the coefficient vector of the original polynomial,

At issue then, at least in the estimation of worst case behaviour, is the relative magnitudes of the vectors **a** and **v**. Since these are truncation errors, their size depends upon the type of truncation and upon the relative sizes of the coefficient vectors. For example, if **A** and **V** were of the same size, one might expect from the above uniform bounds that the error terms associated with the Bézier or B-spline representations might be better behaved than that associated with the explicit polynomial coefficients, because of the factor of $n + 1$ in the first inequality.

In these experiments it was discovered that the coefficients **A** and **V** are not generally of the same size. In fact, the relative sizes of the respective coefficients for each of the three representations can vary quite dramatically. To demonstrate this variation, we display in Figure 19 the polynomial coefficients, and the Bézier and B-spline control points, for the fourth degree polynomial approximation to the unit semi-circle. For the sake of the illustration the consecutive coefficients of the explicit polynomial are joined to one another in a manner similar to the Bézier and B-spline control polygons. One rather striking feature of this figure is the relative sizes of the convex hulls of these coefficients. Clearly, truncation of the larger coefficients will result in a greater loss of information than a corresponding truncation of smaller coefficients. Such a loss was observed in our experiments, and the behaviour was little changed as the polynomial was translated or scaled away from the origin.

It is well understood that the Bézier control polygon lies 'near' the curve which it describes, this being more so as the degree increases. The polynomial coefficients do not generally enjoy this property, exhibiting a more random and less intuitive orientation. As regards the B-spline control points, they share some of the properties of the Bézier control points, such as the convex hull property. It is also known that there is a certain 'laying-off' property associated with the uniform B-splines. That is the curve is not usually as good an approximation of the B-spline control polygon as for the Bézier case.

What is not very well publicized is the fact that for curves of relatively small degree, the uniform B-spline control polygon can be quite far from the curve which it represents. In Figure 20 we illustrate this behaviour by displaying the uniform B-spline control polygon for the polynomial approximations to the unit semi-circle of degree 6.

3.3.3   Results

In our notation the truncated polynomial $\mathbf{TP}(u)$, the reconstituted Bézier polynomial $\mathbf{RTB}(u)$ and the reconstituted B-spline polynomial $\mathbf{RTQ}(u)$ are all images of the original polynomial $\mathbf{P}(u)$. To estimate the respective deviations of these polynomials from the original we use two measures of comparison: the uniform norm and

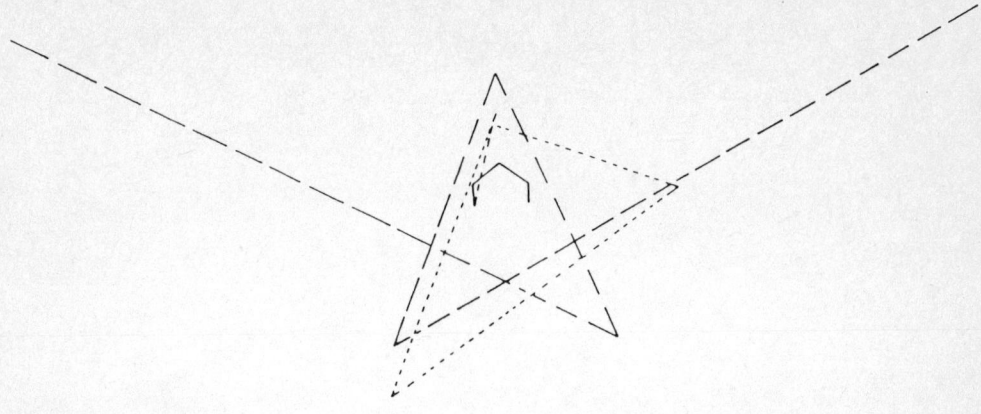

Figure 19. Quartic approximation to unit semi-circle. Solid line - Bézier control polygon, broken line - B-spline control polygon, dotted line - polynomial coefficients.

the least squares norm denoted $\| \ \|_{L_\infty}$ and $\| \ \|_{L_2}$ respectively. The uniform norm estimate is based upon comparing a sample of 100 pairs of points corresponding to equally-spaced parametric values in the interval [0,1], while the least squares estimate is based upon Romberg iterations applied to Simpson's Rule.

For the fourth degree polynomial example illustrated previously, we found that the uniform norms of the respective differences satisfied

$$\|\mathbf{P} - \mathbf{RTB}\|_{L_\infty} \approx 0.1426D - 05,$$

$$\|\mathbf{P} - \mathbf{TP}\|_{L_\infty} \approx 3.51\|\mathbf{P} - \mathbf{RTB}\|_{L_\infty},$$

$$\|\mathbf{P} - \mathbf{RTQ}\|_{L_\infty} \approx 1.34\|\mathbf{P} - \mathbf{RTB}\|_{L_\infty},$$

while the least squares norms satisfied

$$\|\mathbf{P} - \mathbf{RTB}\|_{L_2} \approx 0.1045D - 05,$$

Figure 20. Uniform B-spline control polygon for polynomial approximation of degree 6 to unit semi-circle

$$\|\mathbf{P} - \mathbf{RTB}\|_{L_2} \approx 0.1045D - 05,$$

$$\|\mathbf{P} - \mathbf{TP}\|_{L_2} \approx 2.14\|\mathbf{P} - \mathbf{RTB}\|_{L_2},$$

$$\|\mathbf{P} - \mathbf{RTQ}\|_{L_2} \approx 1.77\|\mathbf{P} - \mathbf{RTB}\|_{L_2}.$$

Note that in this instance the reconstituted truncated Bézier curve $\mathbf{RTB}(u)$ is 'closer' to the original curve $\mathbf{P}(u)$ than either $\mathbf{TP}(u)$ or $\mathbf{RTQ}(u)$, whether measured in the L_∞ or L_2 norms.

The comparisons illustrated above were made for the three types of truncation formats, for polynomials of degrees 2 through 15 approximating the five different planar curves. Table 1 displays the average ratios of the uniform norm comparing truncated explicit polynomial and truncated B-spline with truncated Bézier while Table 2 does the same for the least squares norm.

3.3.4 Conclusions

From these experiments it was concluded that the power series coefficients do not provide the most stable format for the communication of parametric polynomial curves and surfaces. A further disadvantage of this representation when applied to piecewise parametric curves and surfaces is that, unlike control point represen-

	Format F25.4=4 d.p D25.6=digits D25.12=12 digits	Uniform $\frac{\|P-TP\|_\infty}{\|P-RTB\|_\infty}$	Uniform $\frac{\|P-RTQ\|_\infty}{\|P-RTB\|_\infty}$
Semi-circle center: (0,0) radius:	F25.4 D25.6 1 D25.12	4.73 4.81 2.72	1.44 174.96 31.67
Semi-circle center: $(2000,500\pi)$ radius: 1	F25.4 D25.6 D25.12	2.68 0.99 0.82	0.80 8.66 1.55
Semi-circle center: (2000,0) radius: 2000	F25.4 D25.6 D25.12	6.92 14.73 13.88	1.22 128.34 29.76
Exponential - Polynomial Mix	F25.4 D25.6 D25.12	2.21 1.74 0.79	0.61 756.73 1600.31
Unit Square	F25.4 D25.6 D25.12	5.53 1745830.00 4.3382.84	8.36 104206108.30 4.3731.13

Table 1. Average ratio of uniform norms

tations, it cannot guarantee that adjacent sections are indeed coincident at their common ends or edges. For the communication of high degree curves and surfaces the strictly uniform B- spline representation is also less stable than the Bézier representation or B-splines using coincident end knots. Both the CAD*I neutral file specification and STEP provide rational B-spline curve and surface entities which includes a special facility for the communication of Bézier control points. These entities are suitable for the communication of rational or polynomial curve or surface data of any degree.

	Format	Least sq. $\frac{\|P-TP\|_2}{\|P-RTB\|_2}$	Least sq. $\frac{\|P-RTQ\|_2}{\|P-RTB\|_2}$
	F25.4=4 d.p D25.6=digits D25.12=12 digits		
Semi-circle center: (0,0) radius:	F25.4 D25.6 1 D25.12	2.20 3.66 2.12	1.78 185.68 59.17
Semi-circle center: $(2000,500\pi)$ radius: 1	F25.4 D25.6 D25.12	2.10 1.65 1.90	1.10 8.35 3.20
Semi-circle center: (2000,0) radius: 2000	F25.4 D25.6 D25.12	3.27 9.34 8.27	1.42 196.06 45.97
Exponential - Polynomial Mix	F25.4 D25.6 D25.12	1.69 2.10 1.05	1.10 3350.19 8162.71
Unit Square	F25.4 D25.6 D25.12	3.03 705717.00 43731.13	10.92 95622484.00 836727671.41

Table 2. Average ratio of least square norms: $P(u)$ - original polynomial, $TP(u)$ - truncated polynomial, $RTB(u)$ - reconstituted Bézier polynomial, $RTQ(u)$ - reconstituted B-spline polynomial

4 CONVERSIONS BETWEEN REPRESENTATIONS

Principal author: P. A. Sherar

4.0   Introduction

A wide variety of CAD systems are in use today. The mathematical forms of the curve and surface entities within such systems show a similar diversity. This can lead to two types of problem when considering the transfer of such data between two dissimilar systems. First, there may be a basic incompatibilty in the curve and surface description employed, e.g the sending system might use rational polynomials while the receiving system may only support non-rational forms of a specified degree. Second, the two systems may employ curve and surface entities based on the same underlying building blocks, in particular parametric polynomials of a given degree, but differ in the actual representation used. The first of these problems requires effective approximation of the sending system form to suit that required by the receiving one and is expanded on in chapters 5 and 6. The second problem can in principle be solved by an exact conversion between the two representations involved. We shall be concerned with the solution to the second problem here.

The most popular form for the description of curves and surfaces in CAD systems today is that employing parametric polynomials. These are most commonly expressed with respect to one of the following four different representations:

1) Explicit polynomial

2) Bézier-Bernstein

3) B-spline (uniform and non-uniform)

4) Ferguson-Hermite.

Although exact conversions between the above representations are theoretically possible the limitations of finite precision arithmetic means that errors due to rounding are likely to propogate. Therefore it is essential that when conversions are required stable algorithms are used. In this chapter we survey some of the stable algorithms for the conversions between representations 1, 2 and 3. Further details on the numerical stability question and on the condition of the bases involved can be found in chapter 3 and, in particular with regard to the Bézier-Bernstein basis, in [12] and [13].

4.1   Parametrisation

The single curve segment and surface patch entities in whichever of the above representations are most commonly parametrised over $[0,1]$ and $[0,1]*[0,1]$ respectively. The corresponding composite entities commonly possess these parametrisations for the individual segments/patches, but also have, in addition, an associated set of global parameter values describing the curve or surface as a whole. These parameter values will in general have a non-uniform spacing. The continuity level that the curve or surface possesses (typically C^1 or C^2) is understood to be with respect to this global parametrisation. If it is necessary to test a given entity for a particular continuity level, e.g for representation conversion purposes, then the relative spacing of the parameter values of adjacent segments or patches must be taken into account. For more details on this topic we refer the reader to [11].

Many of the formulae for representation conversion described here relate to curve and surface entities that are parametrised locally over $[0,1]$ and $[0,1]*[0,1]$ respectively. It is sometimes required to reparametrise a given segment or surface patch to and from the $[0,1]$ and $[0,1]*[0,1]$ forms. Appendix 2 at the end of the chapter lists some formulae and procedures for performing these conversions when the entities are expressed in explicit polynomial form.

4.2   Degree elevation

Although this chapter is concerned with the 'exact' conversion between the various representations of polynomial curves and surfaces of the same degree, an exact conversion, theoretically at least, can be achieved between a curve or surface of a given degree and one of higher degree. Clearly, for example, a curve segment of degree n in explicit polynomial form

$$\mathbf{P}(u) = \sum_{i=0}^{n} \mathbf{a}_i u^i$$

can be viewed as a curve segment of degree $n+1$ by taking the $n+1\,th$ coefficient to be zero:

$$\mathbf{P}(u) = \sum_{i=0}^{n} \mathbf{a}_i u^i + 0 u^{n+1}.$$

By taking successive coefficients as zero we can obviously view the curve as one of any higher degree.

For curve segments in Bézier form degree elevation is achieved in the following way (see [11]):

$$P(u) = \sum_{i=0}^{n} v_i B_{n,i}(u) = \sum_{i=0}^{n+1} v_i' B_{n+1,i}(u),$$

where

$$v_i' = \frac{i}{n+1} v_{i-1} + \left(1 - \frac{i}{n+1}\right) v_i \quad \text{for} \quad i = 0, \ldots, n+1.$$

By iteration we can elevate the degree of a given Bézier curve to any higher degree using this formula. The corresponding formula for the degree elevation of a B-spline curve is detailed in [19].

Surface degree elevation can be accomplished in the following way. For each representation, explicit polynomial, Bézier or B-spline, consider the coefficients/control points of the surface as a 2D array of points (i.e for x, y and z). First, by considering each row of points as defining a curve, apply the corresponding curve degree elevation algorithm to each row of points in the array. The new surface coefficients/control points are then obtained by applying the curve algorithm once more to each subsequent column of points in this new array.

By employing the above algorithms repeatedly the following routines can be used to convert a curve or surface from one representation to another, where the output degree can be any integer value greater than or equal to that possessed by the input.

4.3 Bézier ⟷ explicit polynomial

4.3.1 Bézier to explicit polynomial: curves

A single segment Bézier curve of degree n is written in the form

$$P(u) = \sum_{i=0}^{n} v_i B_{n,i}(u) \quad \text{for} \quad u \in [0,1], \tag{4.1}$$

where $B_{n,i}(u) = \binom{n}{i}(1-u)^{n-i}u^i$ and $\binom{n}{i} = n!/[(n-i)!i!]$.

The v_i are the Bézier control points and the $B_{n,i}(u)$ the Bézier-Bernstein basis functions. We wish to express $P(u)$ in explicit polynomial form:

$$\mathbf{P}(u) = \sum_{i=0}^{n} \mathbf{a}_i u^i$$

for some coefficients $\mathbf{a}_i$. To do this we write (4.1) in the form

$$\mathbf{P}(u) = (\, B_{n,0}(u) \quad \dots \quad B_{n,n}(u)\,)(\,\mathbf{v}_0 \quad \dots \quad \mathbf{v}_n\,)^T$$

$$= (\,1 \quad u \quad \dots \quad u^n\,)\Big[M_n\Big](\,\mathbf{v}_0 \quad \dots \quad \mathbf{v}_n\,)^T. \tag{4.2}$$

By expanding out the Bézier-Bernstein basis functions $B_{n,i}(u)$ we find that the elements m_{ij} of the conversion matrix $[M_n]$ satisfy the formula [7]

$$m_{ij} = \begin{cases} (-1)^{i-j}\binom{n}{j}\binom{i}{j} & 0 \le j \le i \le n \\ 0 & \text{otherwise} \end{cases}.$$

If we write the explicit polynomial form of $\mathbf{P}(u)$ as

$$\mathbf{P}(u) = (\,1 \quad u \quad \dots \quad u^n\,)(\,\mathbf{a}_0 \quad \dots \quad \mathbf{a}_n\,)^T,$$

and equate this with (4.2) we obtain the formula

$$(\,\mathbf{a}_0 \quad \dots \quad \mathbf{a}_n\,)^T = \Big[M_n\Big](\,\mathbf{v}_0 \quad \dots \quad \mathbf{v}_n\,)^T, \tag{4.3}$$

or equivalently

$$\mathbf{a}_i = \sum_{j=0}^{n} m_{ij}\mathbf{v}_j \quad \text{for} \ i = 0,...,n.$$

Specialising to the cubic case this gives

$$\begin{pmatrix} \mathbf{a}_0 \\ \mathbf{a}_1 \\ \mathbf{a}_2 \\ \mathbf{a}_3 \end{pmatrix} = \begin{pmatrix} 1 & 0 & 0 & 0 \\ -3 & 3 & 0 & 0 \\ 3 & -6 & 3 & 0 \\ -1 & 3 & -3 & 1 \end{pmatrix} \begin{pmatrix} \mathbf{v}_0 \\ \mathbf{v}_1 \\ \mathbf{v}_2 \\ \mathbf{v}_3 \end{pmatrix}.$$

To convert a composite Bézier curve to explicit polynomial form we can simply apply (4.3) to each segment. Note that conversion to explicit polynomial form using the above matrix method will result in each polynomial segment being parametrised locally over $[0,1]$.

4.3.2 Bézier to explicit polynomial: surfaces

If $\mathbf{P}(u,v)$ is a single Bézier surface patch of degree m by n then we can write

$$\mathbf{P}(u,v) = \sum_{i=0}^{m}\sum_{j=0}^{n} \mathbf{v}_{ij}B_{m,i}(u)B_{n,j}(v) \quad \text{for } (u,v) \in [0,1]*[0,1]$$

$$= (B_{m,0}(u) \quad \dots \quad B_{m,m}(u))\Big[\mathbf{v}_{ij}\Big](B_{n,0}(v) \quad \dots \quad B_{n,n}(v))^{T}$$

$$= (1 \quad u \quad \dots \quad u^{m})\Big[M_m\Big]\Big[\mathbf{v}_{ij}\Big]\Big[M_n\Big]^{T}(1 \quad v \quad \dots \quad v^{n})^{T}. \tag{4.4}$$

By writing $\mathbf{P}(u,v)$ in explicit polynomial form,

$$\mathbf{P}(u,v) = \sum_{i=0}^{m}\sum_{j=0}^{n} \mathbf{a}_{ij}u^{i}v^{j}$$

$$= (1 \quad u \quad \dots \quad u^{m})\Big[\mathbf{a}_{ij}\Big](1 \quad v \quad \dots \quad v^{n})^{T}, \tag{4.5}$$

and comparing (4.4) with (4.5) we obtain the formula

$$\Big[\mathbf{a}_{ij}\Big] = \Big[M_m\Big]\Big[\mathbf{v}_{ij}\Big]\Big[M_n\Big]^{T}. \tag{4.6}$$

To convert a composite Bézier surface to explicit polynomial form we can apply (4.6) in a piecewise fashion to each patch. Each individual patch of the resulting polynomial surface will be parametrised locally over $[0,1]*[0,1]$.

4.3.3 Explicit polynomial to Bézier: curves

For the conversion from the explicit polynomial form to the Bézier form we have from equation (4.3)

$$\Big[M_n\Big]^{-1}(\mathbf{a}_0 \quad \dots \quad \mathbf{a}_n)^{T} = (\mathbf{v}_0 \quad \dots \quad \mathbf{v}_n)^{T}.$$

It is not necessary to compute the inverse of $[M_n]$ as an explicit formula for the coefficients of this matrix exists. To see this we factorise $[M_n]$ as follows:

$$\Big[M_n\Big] = \Big[G_n\Big]\Big[H_n\Big],$$

where $\left[G_n\right] = (g_{ij})^n_{i,j=0}$, $\left[H_n\right] = (h_{ij})^n_{i,j=0}$, and

$$g_{ij} = \begin{cases} \binom{n}{j} & i = j \\ 0 & \text{otherwise} \end{cases} \quad \text{and,} \quad h_{ij} = \begin{cases} (-1)^{i-j}\binom{i}{j} & 0 \le j \le i \le n \\ 0 & \text{otherwise} \end{cases}.$$

Using this result we have

$$\left[M_n\right]^{-1} = \left[H_n\right]^{-1}\left[G_n\right]^{-1},$$

where

$$g_{ij}^{-1} = \begin{cases} 1/\binom{n}{j} & i = j \\ 0 & \text{otherwise} \end{cases} \quad \text{and} \quad h_{ij}^{-1} = \begin{cases} \binom{i}{j} & 0 \le j \le i \le n \\ 0 & \text{otherwise} \end{cases}.$$

Hence we can write

$$\mathbf{v}_i = \sum_{j=0}^{n} m_{ij}^{-1}\mathbf{a}_j \quad \text{for } i = 0, ..., n,$$

where

$$m_{ij}^{-1} = \begin{cases} \binom{i}{j}/\binom{n}{j} & 0 \le j \le i \le n \\ 0 & \text{otherwise} \end{cases}.$$

Specialising to the cubic case this gives

$$\begin{pmatrix} \mathbf{v}_0 \\ \mathbf{v}_1 \\ \mathbf{v}_2 \\ \mathbf{v}_3 \end{pmatrix} = \begin{pmatrix} 1 & 0 & 0 & 0 \\ 1 & 1/3 & 0 & 0 \\ 1 & 2/3 & 1/3 & 0 \\ 1 & 1 & 1 & 1 \end{pmatrix} \begin{pmatrix} \mathbf{a}_0 \\ \mathbf{a}_1 \\ \mathbf{a}_2 \\ \mathbf{a}_3 \end{pmatrix}.$$

We can use this matrix method to convert a composite explicit polynomial curve to Bézier form if each original segment is parametrised locally over [0,1].

4.3.4 Explicit polynomial to Bézier: surfaces

To convert a surface patch written in explicit polynomial form to its equivalent Bézier form we rearrange equation (4.6) to obtain

$$\left[\mathbf{v}_{ij}\right] = \left[M_m\right]^{-1}\left[\mathbf{a}_{ij}\right]\left[M_n\right]^{-1T}. \tag{4.7}$$

To convert a composite surface in explicit polynomial form to Bézier we can apply (4.7) to each patch under the assumption that the original patches are parametrised locally over $[0, 1] * [0, 1]$.

4.3.5   Summary

For the conversions between the explicit polynomial and Bézier representations of curve and surface entities a simple process of matrix multiplication will perform the operation required. It is possible either to store the relevant matrices, retrieving them when required, or to generate the terms of the matrix internally at the start of the conversion sequence. Portability considerations might well lead to the latter alternative in order to obtain the maximum accuracy for the particular computing machine in question. If however storage is considered, an alternative to storing all the relevant matrices up to the maximum degree required would be to store just the terms of H_{25} say. Using these values it is relatively easy to generate the terms of $H_n, H_n^{-1}, G_n, G_n^{-1}$ for $n \leq 25$.

4.4   B-spline $\longleftrightarrow$ Bézier

4.4.1   B-spline (uniform) to Bézier: curves

Throughout this section we will denote by $N_{i,k}(u)$ the normalised B-spline basis function of order k (=degree+1) on the knot set $(u_i, \ldots, u_{i+k})$ (also written as $\{u_j\}_{j=i}^{i+k}$).

If $\mathbf{P}(u)$ is a single span B-spline curve of degree n on a uniform knot set $(u_0, \ \cdot \ \cdot \ \cdot \ , u_{2n+1})$, where the knot spacing is equal to one (see Figure 21), then we can write

$$\mathbf{P}(u) = \sum_{i=0}^{n} \mathbf{w}_i N_{i,n+1}(u) \quad \text{for } u \in [0, 1],$$

$$= (1 \quad u \quad \ldots \quad u^n) \begin{bmatrix} U_n \end{bmatrix} (\mathbf{w}_0 \quad \ldots \quad \mathbf{w}_n)^T, \tag{4.8}$$

where the matrix $\begin{bmatrix} U_n \end{bmatrix} = (u_{ij})_{i,j=0}^{n}$ is such that (see [9])

$$u_{ij} = \frac{1}{n!} \binom{n}{i} \sum_{k=j}^{n} (n-k)^i (-1)^{k-j} \binom{n+1}{k-j}.$$

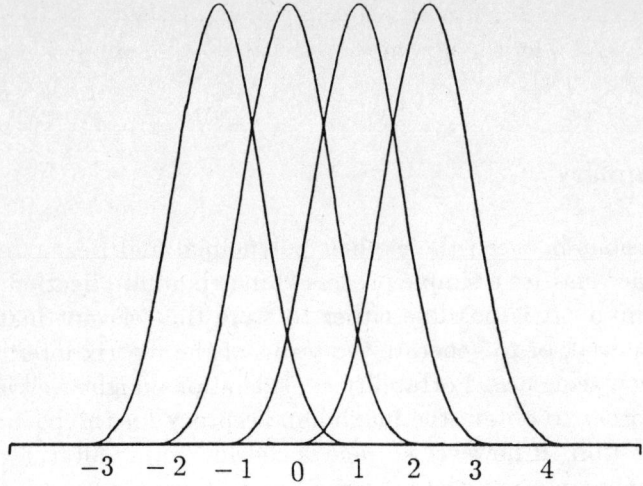

Figure 21: Uniform cubic B-spline basis functions over [0,1]

For the cubic case we have

$$\left[U_3\right] = \frac{1}{6}\begin{pmatrix} 1 & 4 & 1 & 0 \\ -3 & 0 & 3 & 0 \\ 3 & -6 & 3 & 0 \\ -1 & 3 & -3 & 1 \end{pmatrix}.$$

If we write $\mathbf{P}(u)$ in Bézier form,

$$\mathbf{P}(u) = \begin{pmatrix} 1 & u & \cdots & u^n \end{pmatrix}\left[M_n\right]\begin{pmatrix}\mathbf{v}_0 & \cdots & \mathbf{v}_n\end{pmatrix}^T,$$

and compare this with (4.8) we obtain

$$\begin{pmatrix}\mathbf{v}_0 & \cdots & \mathbf{v}_n\end{pmatrix}^T = \left[M_n\right]^{-1}\left[U_n\right]\begin{pmatrix}\mathbf{w}_0 & \cdots & \mathbf{w}_n\end{pmatrix}^T. \qquad (4.9)$$

Specialising to the cubic case this gives

$$\begin{pmatrix}\mathbf{v}_0 \\ \mathbf{v}_1 \\ \mathbf{v}_2 \\ \mathbf{v}_3\end{pmatrix} = \frac{1}{6}\begin{pmatrix} 1 & 0 & 0 & 0 \\ 1 & 1/3 & 0 & 0 \\ 1 & 2/3 & 1/3 & 0 \\ 1 & 1 & 1 & 1 \end{pmatrix}\begin{pmatrix} 1 & 4 & 1 & 0 \\ -3 & 0 & 3 & 0 \\ 3 & -6 & 3 & 0 \\ -1 & 3 & -3 & 1 \end{pmatrix}$$

$$= \frac{1}{6}\begin{pmatrix} 1 & 4 & 1 & 0 \\ 0 & 4 & 2 & 0 \\ 0 & 2 & 4 & 0 \\ 0 & 1 & 4 & 1 \end{pmatrix}\begin{pmatrix}\mathbf{w}_0 \\ \mathbf{w}_1 \\ \mathbf{w}_2 \\ \mathbf{w}_3\end{pmatrix}.$$

To convert a composite B-spline curve on a uniform knot set to Bézier we can apply (4.9) to each segment of the curve. For example, if we need to convert a two segment cubic B-spline curve on a uniform knot set to Bézier form we would apply the matrix transformation to control points $(\mathbf{w}_0, \mathbf{w}_1, \mathbf{w}_2, \mathbf{w}_3)$, and then to the control points $(\mathbf{w}_1, \mathbf{w}_2, \mathbf{w}_3, \mathbf{w}_4)$.

4.4.2   B-spline (uniform) to Bézier: surfaces

For a single B-spline surface patch of degree m by n on a uniform knot set in u and v, $\{u_i\}_{i=0}^{2m+1} \times \{v_j\}_{j=0}^{2n+1}$, we compare the B-spline representation

$$\mathbf{P}(u,v) = (\,1 \quad u \quad \ldots \quad u^m\,) \left[U_m\right] \left[\mathbf{w}_{ij}\right] \left[U_n\right]^T (\,1 \quad v \quad \ldots \quad v^n\,)^T$$

with the equivalent Bézier form

$$\mathbf{P}(u,v) = (\,1 \quad u \quad \ldots \quad u^m\,) \left[M_m\right] \left[\mathbf{v}_{ij}\right] \left[M_n\right]^T (\,1 \quad v \quad \ldots \quad v^n\,)^T,$$

to obtain, on setting $\left[V_n\right] = \left[M_n\right]^{-1} \left[U_n\right]$, the result

$$\left[\mathbf{v}_{ij}\right] = \left[V_m\right] \left[\mathbf{w}_{ij}\right] \left[V_n\right]^T. \tag{4.10}$$

As in the composite B-spline curve case we can apply (4.10) to each patch of a composite uniform B-spline surface in order to generate the equivalent multi-patch Bézier surface. Each Bézier patch will be parametrised locally over $[0, 1] * [0, 1]$.

4.4.3   B-spline (non-uniform) to Bézier: curves

To convert a B-spline curve on a non-uniform knot set to its equivalent Bézier form we must use a more general approach as the B-spline basis functions are no longer simply related to one another. The technique required for conversion of this more general B-spline curve goes under the name of *knot insertion*.

The Bézier-Bernstein basis functions $(B_{n,0}(u), \ldots, B_{n,n}(u))$ for a curve of degree n are seen to be a special case of the B-spline basis functions. In particular, they are the B-spline basis functions on the non-uniform knot set consisting of 0 and 1 when both occur with multiplicity $n+1$. Figure 22 illustrates this for the cubic case. Hence if we are given a B-spline curve $\mathbf{P}(u)$ of order k say,

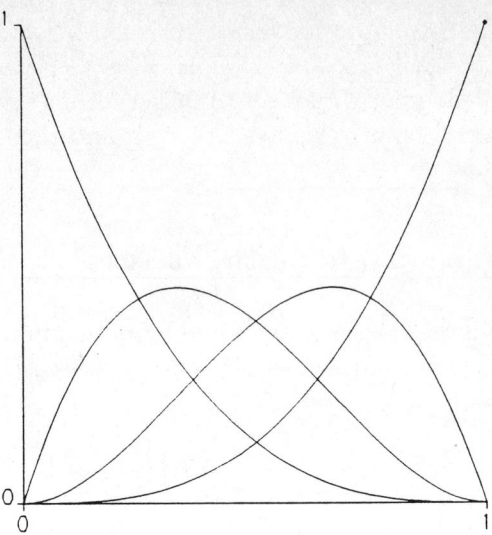

Figure 22. Cubic B-spline basis functions on knot set $(0,0,0,0,1,1,1,1)$=Bézier

$$\mathbf{P}(u) = \sum_{i=0}^{p} \mathbf{w}_i N_{i,k}(u) \quad \text{with knot set } (u_0, \ldots, u_{p+k}), \qquad (4.11)$$

in order to find the Bézier control points defining the same curve we need to generate the Bézier-Bernstein basis functions over the relevant sub-intervals in the sequence $\{u_i\}_{i=0}^{p+k}$. For example, if $\mathbf{P}(u)$ is a three segment B-spline curve on the knot set $(0,0,0,0,1,2,2,3,3,3,3)$,

$$\mathbf{P}(u) = \sum_{i=0}^{6} \mathbf{w}_i N_{i,4}(u),$$

we need to effect the transformation shown in Figure 23. There are two points to note here.

(1) The resulting Bézier curve will in general be composite. In fact if $\mathbf{P}(u)$ is given by (4.11) and the *distinct* knots occurring in the sequence $(u_k, \ldots, u_p)$ are written in order as $(\gamma_1, \ldots, \gamma_l)$ say, with each knot γ_i occurring with multiplicity μ_i in the original set, then $\mathbf{P}(u)$ will have $p - k + 2 - \sum_{i=1}^{l} (\mu_i - 1)$ segments.

(2) To obtain the Bézier control points we need to insert knots into the original set so that each distinct knot in the sequence $(u_{k-1}, \ldots, u_{p+1})$ occurs with multiplicity $k - 1$.

There are two well documented algorithms for carrying out this process of knot insertion: the Oslo algorithm [8] and Boehm's algorithm [1]. Both of these proce-

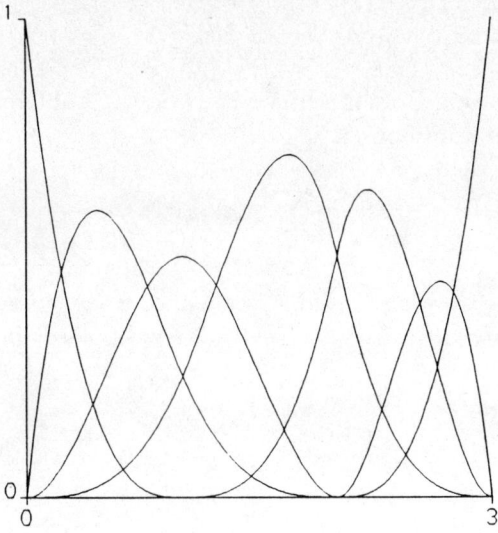

Figure 23a. Cubic B-spline basis functions on knot set (0,0,0,0,1,2,2,3,3,3,3)

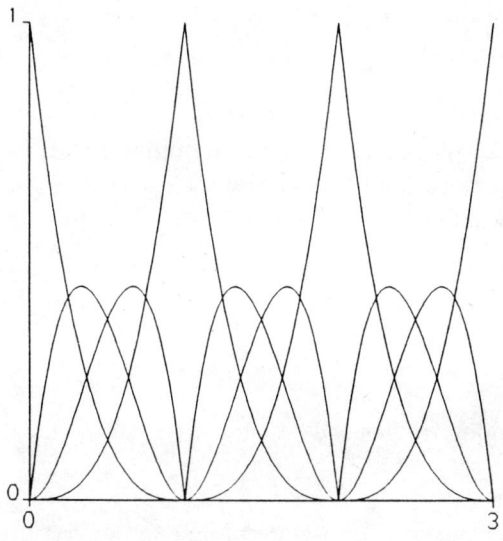

Figure 23b. Cubic B-spline basis functions on knot set (0,0,0,0,1,1,1,2,2,2,3,3,3,3)

dures allow one to add knots to an existing B-spline curve (and more generally a surface) and so generate a new sets of control points in such a way that the resulting curve is identical to the original. There are many useful applications of this process and we refer the reader to [8] for a description of these. For our purposes, that of generating the Bézier control points of B-spline curves and surfaces, we concentrate on Boehm's algorithm in view of its simplicity.

4.4.4 Single knot insertion

Assume then that we have a B-spline curve $\mathbf{P}(u)$ on the knot set $\pi = \{u_i\}$ (for ease of notation we omit the limits on i),

$$\mathbf{P}(u) = \sum_i \mathbf{w}_i N_{i,k}(u).$$

To set the scene suppose we wish to add a single knot $\hat{u}$ say, coinciding with the knot u_{l+1} which already has multiplicity s. Let the new knot set be denoted $\hat{\pi} = \{\hat{u}_i\}$ so that

$$\hat{u}_i = \begin{cases} u_i & i \leq l \\ \hat{u} = u_{l+1} & i = l+1 \\ u_{i-1} & i \geq l+2 \end{cases}.$$

Denoting the new set of basis functions as $\{\hat{N}_{i,k}(u)\}$ we can express $\mathbf{P}(u)$ in the form

$$\mathbf{P}(u) = \sum_i \hat{\mathbf{w}}_i \hat{N}_{i,k}(u), \tag{4.12}$$

for some yet to be determined control points $\hat{\mathbf{w}}_i$. To find these new control points we begin by expressing $N_{i,k}(u)$ as a linear combination of $\hat{N}_{i,k}(u)$ and $\hat{N}_{i+1,k}(u)$ (for derivation see Appendix 1 at end of chapter):

$$N_{i,k}(u) = \begin{cases} \hat{N}_{i,k}(u) \\ (\hat{u} - \hat{u}_i)/(\hat{u}_{i+k} - \hat{u}_i)\hat{N}_{i,k}(u) + (\hat{u}_{i+k+1} - \hat{u})/(\hat{u}_{i+k+1} - \hat{u}_{i+1})\hat{N}_{i+1,k}(u) \\ \hat{N}_{i+1,k} \end{cases}$$

$$\tag{4.13}$$

where the respective index ranges are

$$\begin{cases} i \leq l-k+s \\ l-k+s+1 \leq i \leq l \\ i \geq l+1 \end{cases}.$$

If we now sum $\mathbf{w}_i N_{i,k}(u)$ over the range $l - k + s + 1 \leq i \leq l$ using the above expression, and change the variable of summation ($i \to i-1$) for the terms involving $\hat{N}_{i+1,k}(u)$, we obtain (remembering that $\hat{u} - \hat{u}_{l+1} = 0$),

$$\sum_{i=l-k+s+1}^{l} \mathbf{w}_i N_{i,k}(u) = \sum_{i=l-k+s+1}^{l+1} \left(\frac{\hat{u} - \hat{u}_i}{\hat{u}_{i+k} - \hat{u}_i} \mathbf{w}_i + \frac{\hat{u}_{i+k} - \hat{u}}{\hat{u}_{i+k} - \hat{u}_i} \mathbf{w}_{i-1} \right) \hat{N}_{i,k}(u). \tag{4.14}$$

Hence, by the uniqueness of the representation of $\mathbf{P}(u)$ with respect to a given B-spline basis we have, by comparing (4.12) with (4.14), the result

$$\hat{\mathbf{w}}_i = \alpha_i \mathbf{w}_i + (1 - \alpha_i)\mathbf{w}_{i-1}, \tag{4.15}$$

where

$$\alpha_i = \begin{cases} 1 & i \leq l - k + s + 1 \\ (\hat{u} - \hat{u}_i)/(\hat{u}_{i+k} - \hat{u}_i) = (\hat{u} - u_i)/(u_{i+k-1} - u_i) & l - k + s + 2 \leq i \leq l \\ 0 & i \geq l + 1 \end{cases}. \tag{4.16}$$

Equation (4.15) has the simple interpretation of $\hat{\mathbf{w}}_i$ dividing the line joining $\mathbf{w}_{i-1}$ and $\mathbf{w}_i$ in the ratio $\alpha_i : 1 - \alpha_i$. The calculation of the new set of control points involves only *convex* combinations of the original control points and hence the process is computationally stable.

Having found a formula for the new control points after a single knot insertion we note that in general to obtain the Bézier representation we will have to insert a particular knot more than once. We examine the process needed to achieve this.

4.4.5   Multiple knot insertion

To insert the knot $\hat{u} = u_{l+1}$ with multiplicity r say (where $r + s \leq k - 1$), we apply the above process (4.4.4) of single knot insertion iteratively. If we denote the resulting knot set as $\{u_i^r\}$ so that

$$u_i^r = \begin{cases} u_i & i \leq l \\ \hat{u} = u_{l+1} & l + 1 \leq i \leq l + r \\ u_{i-r} & l + r + 1 \leq i \end{cases},$$

and the corresponding expression for $\mathbf{P}(u)$ by

$$\mathbf{P}(u) = \sum_i \mathbf{w}_i^r N_{i,k}^r(u),$$

then assuming that $\hat{u}$ has been inserted $r - 1$ times, we have, according to (4.15) and (4.16)

$$\mathbf{w}_i^r = \alpha_i^r \mathbf{w}_i^{r-1} + (1 - \alpha_i^r)\mathbf{w}_{i-1}^{r-1} \quad, \mathbf{w}_i^0 = \mathbf{w}_i, \tag{4.17}$$

where

$$\alpha_i^r = \begin{cases} 1 & i \leq l - k + s + r \\ (\hat{u} - u_i^r)/(u_{i+k}^r - u_i^r) = (\hat{u} - u_i)/(u_{i+k-r} - u_i) & l - k + s + r + 1 \leq i \leq l \\ 0 & i \geq l + 1 \end{cases}.$$

There are some important points to note arising from the above formulae.

(1) Because $\alpha_i^r = 1$ for $i \leq l - k + s + r$, the points $\mathbf{w}_i^j$ coincide for $i =$ constant, $i - j \leq l - k + s + 1$.

(2) Because $\alpha_i^r = 0$ for $i \geq l + 1$, the points $\mathbf{w}_i^j$ coincide for $i - j = $ constant, $i \geq l$.

(3) Since $N_{i,k}^{k-1-s}(\hat{u}) = \delta_{il}$ (1 if $i = l$, 0 otherwise) for $r = k - 1 - s$ we have

$$\mathbf{P}(\hat{u}) = \sum_i \mathbf{w}_i^r N_{i,k}^r(\hat{u}) = \mathbf{w}_l^{k-1-s}.$$

Example 1

Let $\mathbf{P}(u) = \sum_{i=0}^{6} \mathbf{w}_i N_{i,4}(u)$ on the knot set $(0,0,0,0,1,2,2,3,3,3,3)$. We illustrate the process of inserting $\hat{u} = 1$ twice (here $l = 3$ and $s = 1$), see Figure 24.

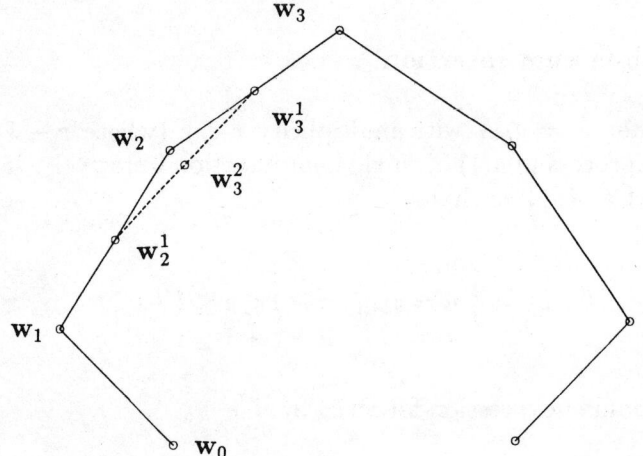

Figure 24. Knot insertion example

Step 1: $\mathbf{P}(u) = \sum_{i=0}^{7} \mathbf{w}_i^1 N_{i,4}^1(u)$ on the knot set $(0,0,0,0,1,1,2,2,3,3,3,3)$, where $(\mathbf{w}_0^1 \ \ldots \ \mathbf{w}_7^1) = (\mathbf{w}_0 \ \ \mathbf{w}_1 \ \ \mathbf{w}_2^1 \ \ \mathbf{w}_3^1 \ \ \mathbf{w}_3 \ \ \mathbf{w}_4 \ \ \mathbf{w}_5 \ \ \mathbf{w}_6)$.

and

$$\mathbf{w}_2^1 = \frac{1}{2}\mathbf{w}_2 + \frac{1}{2}\mathbf{w}_1, \qquad \mathbf{w}_3^1 = \frac{1}{2}\mathbf{w}_3 + \frac{1}{2}\mathbf{w}_2.$$

Step 2: $\mathbf{P}(u) = \sum\limits_{i=0}^{8} \mathbf{w}_i^2 N_{i,4}^2(u)$ on the knot set $(0,0,0,0,1,1,1,2,2,3,3,3,3,3)$, where $(\, \mathbf{w}_0^2 \quad \cdots \quad \mathbf{w}_8^2 \,) = (\, \mathbf{w}_0 \quad \mathbf{w}_1 \quad \mathbf{w}_2^1 \quad \mathbf{w}_3^2 \quad \mathbf{w}_3^1 \quad \mathbf{w}_3 \quad \mathbf{w}_4 \quad \mathbf{w}_5 \quad \mathbf{w}_6 \,)$.

and

$$\mathbf{w}_3^2 = \frac{1}{2}\mathbf{w}_3^1 + \frac{1}{2}\mathbf{w}_2^1.$$

4.4.6 The Cox-de Boor recursion formula

For $l - k + s + r + 1 \leq i \leq l$, (4.15) coincides with the well known Cox-de Boor recursion formula ([5] and [10]):

$$\mathbf{P}(u) = \sum_i \mathbf{w}_i^r N_{i,k-r}(u)$$

$$\mathbf{w}_i^r = \begin{cases} \mathbf{w}_i & i = 0 \\ (u - u_i)/(u_{i+k-r} - u_i)\mathbf{w}_i^{r-1} + (u_{i+k-r} - u)/(u_{i+k-r} - u_i)\mathbf{w}_{i-1}^{r-1} & r > 0 \end{cases}.$$

This formula expresses $\mathbf{P}(u)$ as a linear combination of B-splines of order $k - r$ over the knot set π with corresponding triangular scheme for the calculation of the $\mathbf{w}_i^r$ as shown in the following table:

$\mathbf{w}_{l-k+1}^0$

$\mathbf{w}_{l-k+2}^0 \qquad \mathbf{w}_{l-k+2}^1$

$\cdot \qquad\qquad\qquad \cdot$

$\cdot \qquad\qquad\qquad\qquad\qquad \cdot$

$\cdot \qquad\qquad\qquad\qquad\qquad\qquad\qquad \cdot$

$\cdot \qquad\qquad\qquad\qquad\qquad\qquad\qquad\qquad\qquad \cdot$

$\mathbf{w}_{l-1}^0 \qquad \mathbf{w}_{l-1}^1 \qquad \cdot \qquad \cdot \qquad \cdot \qquad \mathbf{w}_{l-1}^{k-2}$

$\mathbf{w}_l^0 \qquad \mathbf{w}_l^1 \qquad \cdot \qquad \cdot \qquad \cdot \qquad \mathbf{w}_l^{k-2} \qquad \mathbf{w}_l^{k-1}$

The above scheme for calculating the points $\mathbf{w}_i^j$ can be reduced somewhat for our purposes due to the coincidences stated in (1) and (2) above. If u_{l+1} has multiplicity s in the knot set π, we only have to insert $\hat{u}$ $k-1-s$ times to obtain the multiplicity $k-1$ we require. Letting $q = k-1-s$, the reduced Cox de-Boor triangular table looks like the following:

$$\mathbf{w}_{l-q}^0$$

$$\mathbf{w}_{l-q+1}^0 \qquad \mathbf{w}_{l-q+1}^1$$

$$\mathbf{w}_l^0 \qquad \mathbf{w}_l^1 \quad \cdot \quad \cdot \quad \cdot \quad \mathbf{w}_l^q$$

This table contains both the control points that need replacing and their replacements as indicated in the following diagram:

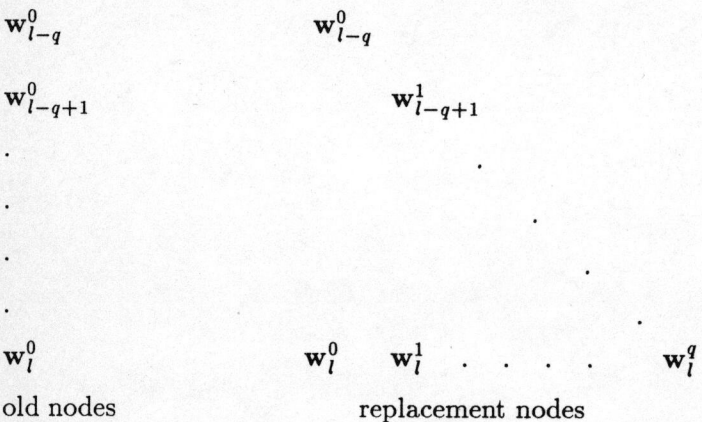

old nodes   replacement nodes

Returning to Example 1 where we insert $\hat{u}$ twice, the original triangular scheme appears as:

$$\mathbf{w}_0$$

$\mathbf{w}_1$	$\mathbf{w}_1^1$		
$\mathbf{w}_2$	$\mathbf{w}_2^1$	$\mathbf{w}_2^2$	
$\mathbf{w}_3$	$\mathbf{w}_3^1$	$\mathbf{w}_3^2$	$\mathbf{w}_3^3$

and the reduced scheme as:

$\mathbf{w}_1$		
$\mathbf{w}_2$	$\mathbf{w}_2^1$	
$\mathbf{w}_3$	$\mathbf{w}_3^1$	$\mathbf{w}_3^2$

This gives us replacement nodes ($\mathbf{w}_1$ $\mathbf{w}_2^1$ $\mathbf{w}_3^2$ $\mathbf{w}_3^1$ $\mathbf{w}_3$) as obtained earlier.

4.4.7 The B-spline to Bézier algorithm

Following the above analysis, the full algorithm for generating the Bézier points of a B-spline curve is given in three steps:

Successively for $u_l < u_{l+1} = u_{k-1}, \ldots, u_l < u_{l+1} = u_{p+1}$. If u_{l+1} is of multiplicity s with $s < k - 1$:

Step 1: Calculate $\mathbf{w}_l^q(u_{l+1})$ using the Cox de-Boor recursion formula where $q = k - 1 - s$.

Step 2: Increase m and for $j > l$ the subscripts of $\mathbf{w}_j$ and u_j by q.

Step 3: For $i = 1, \ldots, q$ replace

u_{l+i} by u_{l+1},

$\mathbf{w}_{l-q+i}$ by $\mathbf{w}_{l-q+i}^i$ generating the new control points $\mathbf{w}_{l-q+1}^1, \ldots, \mathbf{w}_l^q$,

$\mathbf{w}_{l+i}$ by $\mathbf{w}_l^{q-i}$ generating the new control points $\mathbf{w}_l^{q-1}, \ldots, \mathbf{w}_l^0$.

4.4.8 The linear transformation method

As an alternative to the above method we can interpret the Bézier control points as arising from a matrix transformation applied to the original B-spline points.

Example 1 with $\mathbf{P}(u) = \sum_{i=0}^{6} \mathbf{w}_i N_{i,4}(u)$ on the knot set $(0, 0, 0, 0, 1, 2, 2, 3, 3, 3, 3)$

serves to illustrate this approach. Inserting $\hat{u} = 1$ to obtain the knot set $(0, 0, 0, 0, 1, 1, 2, 2, 3, 3, 3, 3)$ we can write

$$(\mathbf{w}_0^1 \quad \cdots \quad \mathbf{w}_7^1)^T = \begin{bmatrix} C_1 \end{bmatrix} (\mathbf{w}_0 \quad \cdots \quad \mathbf{w}_6)^T ,$$

where $\begin{bmatrix} C_1 \end{bmatrix}$ of order 8 by 7 is

$$\begin{pmatrix} 1 & & & & & & \\ & 1 & & & & & \\ & 1 - \alpha_2^1 & \alpha_2^1 & & & & \\ & & 1 - \alpha_3^1 & \alpha_3^1 & & & \\ & & & 1 & & & \\ & & & & 1 & & \\ & & & & & 1 & \\ & & & & & & 1 \end{pmatrix} .$$

Inserting $\hat{u} = 1$ again to give the knot set $(0, 0, 0, 0, 1, 1, 1, 2, 2, 3, 3, 3, 3)$ we can write

$$(\mathbf{w}_0^2 \quad \cdots \quad \mathbf{w}_8^2)^T = \begin{bmatrix} C_2 \end{bmatrix} (\mathbf{w}_0^1 \quad \cdots \quad \mathbf{w}_7^1)^T ,$$

where $\begin{bmatrix} C_2 \end{bmatrix}$ of order 9 by 8 is

$$\begin{pmatrix} 1 & & & & & & & \\ & 1 & & & & & & \\ & & 1 & & & & & \\ & & 1 - \alpha_3^2 & \alpha_3^2 & & & & \\ & & & 1 & & & & \\ & & & & 1 & & & \\ & & & & & 1 & & \\ & & & & & & 1 & \\ & & & & & & & 1 \end{pmatrix} .$$

Finally, inserting $\hat{u} = 2$ to give $(0, 0, 0, 0, 1, 1, 1, 2, 2, 2, 3, 3, 3, 3)$ we have

$$(\mathbf{w}_0^3 \quad \cdots \quad \mathbf{w}_9^3)^T = \begin{bmatrix} C_3 \end{bmatrix} (\mathbf{w}_0^2 \quad \cdots \quad \mathbf{w}_8^2)^T ,$$

where $\begin{bmatrix} C_3 \end{bmatrix}$ of order 10 by 9 is the matrix

$$\begin{pmatrix} 1 & & & & & & & & \\ & 1 & & & & & & & \\ & & 1 & & & & & & \\ & & & 1 & & & & & \\ & & & & 1 & & & & \\ & & & & 1-\alpha_6^3 & \alpha_6^3 & & & \\ & & & & & 1 & & & \\ & & & & & & 1 & & \\ & & & & & & & 1 & \\ & & & & & & & & 1 \end{pmatrix}.$$

By composing the above matrix transformations we have

$$(\mathbf{w}_0^3 \quad \cdots \quad \mathbf{w}_9^3)^T = \begin{bmatrix} C \end{bmatrix} (\mathbf{w}_0 \quad \cdots \quad \mathbf{w}_6)^T ,$$

where $\begin{bmatrix} C \end{bmatrix} = \begin{bmatrix} C_3 \end{bmatrix}\begin{bmatrix} C_2 \end{bmatrix}\begin{bmatrix} C_1 \end{bmatrix}$. More generally, if

$$\mathbf{P}(u) = \sum_{i=0}^{p} \mathbf{w}_i N_{i,k}(u)$$

and our required final representation is written as

$$\mathbf{P}(u) = \sum_{i=0}^{p_1} \mathbf{w}_i^\dagger N_{i,k}^\dagger(u),$$

where $p_1 = p+$ the number of knots inserted, then

$$\mathbf{w}^\dagger = \begin{bmatrix} C \end{bmatrix}\mathbf{w},$$

where $\mathbf{w} = (\mathbf{w}_0 \quad \cdots \quad \mathbf{w}_p)$ and $\mathbf{w}^\dagger = (\mathbf{w}_0^\dagger \quad \cdots \quad \mathbf{w}_{p_1}^\dagger)$.

Here $[C]$ will be a matrix of order $p_1 +1$ by $p+1$ and is equal to the product of the $p_1 - p$ matrices $[C_i]$, $i = 1,\ldots,p_1 - p$, each $[C_i]$ being of order $p+1+i$ by $p+i$. The matrix $[C]$ has the following properties:

(1) The sum of the elements in any row is equal to one (since this is true for each of the matrices $[C_i]$).

(2) There are at most k non-zero elements in each row (due to the local non-zero nature of the basis functions).

(3) The relationship between the original and final set of B-spline basis functions is given by

$$(N_{0,k}(u) \quad \ldots \quad N_{p,k}(u)) = (N^{\dagger}_{0,k}(u) \quad \ldots \quad N^{\dagger}_{k,p_1}(u)) \begin{bmatrix} C \end{bmatrix}. \tag{4.18}$$

Statement (3) follows since

$$\mathbf{P}(u) = \sum_{i=0}^{p_1} \mathbf{w}^{\dagger}_i N^{\dagger}_{i,k}(u) = \sum_{i=0}^{p_1} \sum_{j=0}^{p} c_{ij} \mathbf{w}_j N^{\dagger}_{i,k}(u)$$

$$= \sum_{i=0}^{p} \mathbf{w}_i N_{i,k}(u).$$

If we let $\mathbf{w}_i = (0,0,0)$, $i \neq \alpha$ and $\mathbf{w}_\alpha = (1,0,0)$ say, we obtain

$$N_{\alpha,k}(u) = \sum_{i=0}^{p_1} c_{i,\alpha} N^{\dagger}_{i,k}(u)$$

as required.

This matrix method for converting from B-spline to Bézier will be beneficial when one has to perform the same knot insertion a number of times e.g the surface case described below. The Oslo algorithm generates the non-zero elements of $[C]$, the most efficient form of which is described in [18].

4.4.9   B-spline (non-uniform) to Bézier: surfaces

We can use the B-spline to Bézier curve algorithm (4.4.7) to convert a B-spline surface to its equivalent Bézier form. Let $\mathbf{P}(u,v)$ represent a B-spline surface of order k by l on the knot set $\{u_i\}_{i=0}^{p+k} * \{v_j\}_{j=0}^{q+l}$:

$$\mathbf{P}(u,v) = \sum_{i=0}^{p} \sum_{j=0}^{q} \mathbf{w}_{ij} N_{i,k}(u) N_{j,l}(v). \tag{4.19}$$

The conversion of $\mathbf{P}$ to Bézier form is achieved in the following two steps:

Step 1: Apply the curve algorithm to each of the $p+1$ rows of control points making up the control polygon using the v knot set,

$$\mathbf{P}_i(v) = \sum_{j=0}^{q} \mathbf{w}_{ij} N_{j,l}(v) \quad \text{for } i = 0, 1, \ldots, p.$$

This gives us a new set of control points, call them $\{\mathbf{w}'_{ij}\}_{i=0,j=0}^{p,q_1}$, where $q_1 = q+$ the number of knots inserted into the v knotline, and identical surface

$$\mathbf{P}(u,v) = \sum_{i=0}^{p} \sum_{j=0}^{q_1} \mathbf{w}'_{ij} N_{i,k}(u) N_{j,l}^{\dagger}(v),$$

on the knot set $\{u_i\}_{i=0}^{p+k} * \{v_j^{\dagger}\}_{j=0}^{q_1+l}$, where $\{v_j^{\dagger}\}_{j=0}^{q_1+l}$ is the new v knot set.

Step 2: Apply the curve algorithm to the $q_1 + 1$ columns of this new control net using the u knot set,

$$\mathbf{P}_j(u) = \sum_{i=0}^{p} \mathbf{w}'_{ij} N_{i,k}(u) \quad \text{for } j = 0, 1, \ldots, q_1.$$

The control points obtained from this second step, call them $\{\mathbf{w}_{ij}^{\dagger}\}_{i=0,j=0}^{p_1,q_1}$, are the required Bézier points. The final surface representation is

$$\mathbf{P}(u,v) = \sum_{i=0}^{p_1} \sum_{j=0}^{q_1} \mathbf{w}_{ij}^{\dagger} N_{i,k}^{\dagger}(u) N_{j,l}^{\dagger}(v), \tag{4.20}$$

with associated knot set $\{u_i^{\dagger}\}_{i=0}^{p_1+k} * \{v_j^{\dagger}\}_{j=0}^{q_1+l}$. Here $\{u_i^{\dagger}\}_{i=0}^{p_1+k}$ is the new u knot set and $p_1 = p+$ the number of knots inserted into the u knotline. Finally we note that we could if we wish treat the columns of control points first and then the rows, the result is the same.

4.4.10 The matrix method

We can as in the curve case (4.4.8) express the B-spline to Bézier surface conversion in terms of a matrix transformation $[C]$. In particular, if $[C_{row}]$ is the matrix corresponding to the knot insertion of the u knotline and similarly if $[C_{col}]$ is the corresponding matrix for the v knotline we have by (4.18) the results:

$$(N_{0,l}(v) \quad \cdots \quad N_{q,l}(v)) = (N_{0,l}^{\dagger}(v) \quad \cdots \quad N_{q_1,l}^{\dagger}(v)) \left[C_{col} \right],$$

$$(N_{0,k}(u) \quad \cdots \quad N_{p,k}(u)) = (N_{0,k}^{\dagger}(u) \quad \cdots \quad N_{p_1,k}^{\dagger}(u)) \left[C_{row} \right].$$

Substituting these expressions into the matrix form of (4.19):

$$\mathbf{P}(u,v) = (\, N_{0,k}(u) \quad \ldots \quad N_{p,k}(u)\,) \left[\mathbf{w}_{ij}\right] (\, N_{0,l}(v) \quad \ldots \quad N_{q,l}(v)\,)^T,$$

and comparing the result with the matrix form of (4.20):

$$\mathbf{P}(u,v) = \left(\, N_{0,k}^{\dagger}(u) \quad \ldots \quad N_{p_1,k}^{\dagger}(u)\,\right) \left[\mathbf{w}_{ij}^{\dagger}\right] \left(\, N_{0,l}^{\dagger}(u) \quad \ldots \quad N_{q_1,l}^{\dagger}(v)\,\right)^T$$

gives us the matrix equation

$$\left[\mathbf{w}_{ij}^{\dagger}\right] = \left[C_{row}\right]\left[\mathbf{w}_{ij}\right]\left[C_{col}\right]^T.$$

4.4.11 Summary

We can separate the B-spline to Bézier conversion into two cases; the uniform and non-uniform. For the uniform conversion a simple process of matrix multiplication applied to the B-spline control points will produce the Bézier points in both the curve and surface case. The multiplications can be applied in a piecewise fashion to convert a composite uniform B-spline curve or surface to Bézier form. For the non-uniform case we can use the process of knot insertion to generate the Bézier control points in both the curve and surface case.

4.4.12 Bézier to B-spline: curves

We begin by noting that the Bézier to B-spline conversion is not, theoretically at least, strictly necessary as Bézier curves form a subset of the set of B-spline curves. In particular, for a curve of degree n the internal knots occur with multiplicity n and the two end knots with multiplicity $n+1$. Clearly no computation is needed in order to communicate Bézier data as B-spline data if this approach is taken. Note however that no explicit continuity information is supplied in this case, all that can be deduced is that the resulting curve is C^0 continuous. If a higher continuity level is known to exist methods for converting the Bézier form to B-spline that convey this information can be used. At the expense in general of some computation and hence the possibility of rounding errors there will be a subsequent reduction in the size of the data file produced as the control point set will be smaller.

Consider first a single segment Bézier curve of degree n. From (4.9) we have the result

$$(\mathbf{w}_0 \quad \ldots \quad \mathbf{w}_n)^T = \left[U_n \right]^{-1} \left[M_n \right] (\mathbf{v}_0 \quad \ldots \quad \mathbf{v}_n)^T ,$$

where the single segment B-spline curve with control points $\mathbf{w}_0, \ldots, \mathbf{w}_n$ is based on a uniform integer knot set and is parametrised over $[0, 1]$. To the author's knowledge no explicit formula exists for the elements of $[U_n]^{-1}$ and hence a Gaussian elimination technique is needed to compute the inverse. For the cubic case however we do have the following result:

$$\left[U_3 \right]^{-1} \left[M_3 \right] = \frac{1}{3} \begin{pmatrix} 3 & -3 & 2 & 0 \\ 3 & 0 & -1 & 0 \\ 3 & 3 & 2 & 0 \\ 3 & 6 & 11 & 18 \end{pmatrix} \begin{pmatrix} 1 & 0 & 0 & 0 \\ -3 & 3 & 0 & 0 \\ 3 & -6 & 3 & 0 \\ -1 & 3 & -3 & 1 \end{pmatrix}$$

$$= \begin{pmatrix} 6 & -7 & 2 & 0 \\ 0 & 2 & -1 & 0 \\ 0 & -1 & 2 & 0 \\ 0 & 2 & -7 & 6 \end{pmatrix} .$$

Whether it is possible to use this matrix method to convert a composite Bézier curve to B-spline form depends, as indicated above, on the continuity level of the curve. Concentrating on the cubic case, if the Bézier curve is C^2 continuous we can use the above matrix to compute the first four B-spline control points and then note that for all succeeding segments we need only compute one extra control point,

$$\mathbf{w}_4 = 2\mathbf{v}_1 - 7\mathbf{v}_2 + 6\mathbf{v}_3 \quad \text{etc,}$$

since three control points are common to each pair of neighbouring segments.

If the composite Bézier curve is C^1 continuous we can convert it to 'genuine' B-spline form by simply leaving out the Bézier control points corresponding to the segment joins and modifying the knot set appropriately (this applies to Bézier curves of any degree). We illustrate this in the following example.

Example 2

Suppose we have a three segment cubic Bézier curve with C^1 continuity on the knot set $(0, 0, 0, 0, 1, 1, 1, 2, 2, 2, 3, 3, 3, 3)$. If the original control point set is $(\mathbf{v}_0 \quad \ldots \quad \mathbf{v}_9)$ (so the points corresponding to the segment joins are not repeated), then the equivalent B-spline representation is given by control points $(\mathbf{v}_0 \quad \mathbf{v}_1 \quad \mathbf{v}_2 \quad \mathbf{v}_4 \quad \mathbf{v}_5 \quad \mathbf{v}_7 \quad \mathbf{v}_8 \quad \mathbf{v}_9)$ on the knot set $(0, 0, 0, 1, 1, 2, 2, 3, 3, 3, 3)$. That this is the correct representation can be seen by considering the reverse process of knot insertion. Given a C^1 continuous B-spline curve we need insert only one knot at each segment breakpoint in order to obtain the equivalent Bézier form. For each knot inserted the extra control point generated must lie at the corresponding geometric breakpoint. All other control points remain unaltered.

4.4.13   Bézier to B-spline: surfaces

The remarks concerning the conversion of Bézier curves apply equally well to the surface case. A composite Bézier surface can be interpreted as a B-spline surface on the appropriate knot set in u and v. If however C^1 or C^2 continuity is known to exist, savings can be made by converting to a 'genuine' B-spline.

For a single surface patch of degree m by n we have from (4.10)

$$\left[\mathbf{w}_{ij}\right] = \left[V_m\right]^{-1}\left[\mathbf{v}_{ij}\right]\left[V_n\right]^{-1T}. \tag{4.21}$$

The corresponding B-spline patch will have a uniform knot set in u and v and will be parametrised over $[0,1] * [0,1]$.

To convert a composite C^2 continuous Bézier surface to B-spline we can apply (4.21) in a piecewise fashion. Savings in computation can be made by noting that $m-1$ rows ($n-1$ columns) of control points are common to each pair of neighbouring patches.

If the composite Bézier surface is known to be C^1 we can convert it to B-spline form in two steps:

Step 1: Apply the C^1 Bézier to B-spline curve algorithm to each row of control points in the control net (i.e treat each row as a Bézier curve). This will result in a control net with $p-1$ less control points in each row where p is the number of patches in the v direction.

Step 2: Apply the curve construction to each column of the new control net from Step 1. This will result in a reduction of $q-1$ control points in each column where q is the number of patches in the u direction.

Finally, the knot sets in u and v are modified as in the curve case, each internal knot in u should now occur with multiplicity $m-1$ and those in v with multiplicity $n-1$. Note that the same result is obtained by processing the columns first and then the rows.

4.5 B-spline $\longleftrightarrow$ explicit polynomial

4.5.1 B-spline (uniform) to explicit polynomial: curves

If $\mathbf{P}(u)$ is a single span B-spline curve of degree n on a uniform knot set $(u_0, \ldots, u_{2n+1})$,

$$\mathbf{P}(u) = \sum_{i=0}^{n} \mathbf{w}_i N_{i,n+1}(u) \quad \text{for } u \in [0,1],$$

then from (4.8)

$$\mathbf{P}(u) = (1 \quad u \quad \ldots \quad u^n) \left[U_n \right] (\mathbf{w}_0 \quad \ldots \quad \mathbf{w}_n)^T. \tag{4.22}$$

If we compare this with the explicit polynomial representation

$$\mathbf{P}(u) = (1 \quad u \quad \ldots \quad u^n)(\mathbf{a}_0 \quad \ldots \quad \mathbf{a}_n)^T$$

we obtain

$$(\mathbf{a}_0 \quad \ldots \quad \mathbf{a}_n)^T = \left[U_n \right] (\mathbf{w}_0 \quad \ldots \quad \mathbf{w}_n)^T. \tag{4.23}$$

We can convert a composite B-spline curve on a uniform knot set to explicit polynomial form by applying (4.23) to each segment. Each polynomial segment will be parametrised locally over $[0,1]$.

4.5.2 B-spline (uniform) to explicit polynomial: surfaces

For a single surface patch of degree m by n we compare the B-spline representation

$$\mathbf{P}(u,v) = \sum_{i=0}^{m} \sum_{j=0}^{n} \mathbf{w}_{ij} N_{i,m+1}(u) N_{j,n+1}(v) \quad \text{for } (u,v) \in [0,1] * [0,1]$$

$$= (1 \quad u \quad \ldots \quad u^m) \left[U_m \right] \left[\mathbf{w}_{ij} \right] \left[U_n \right]^T (1 \quad v \quad \ldots \quad v^n)^T,$$

on a uniform knot set $\{u_i\}_{i=0}^{2m+1} * \{v_j\}_{j=0}^{2n+1}$, with the explicit polynomial form

$$\mathbf{P}(u,v) = (1 \quad u \quad \ldots \quad u^m) \left[\mathbf{a}_{ij} \right] (1 \quad v \quad \ldots \quad v^n)^T,$$

to obtain the result

$$\left[\mathbf{a}_{ij}\right] = \left[U_m\right]\left[\mathbf{w}_{ij}\right]\left[U_n\right]^T. \tag{4.24}$$

Formula (4.24) can be applied in a piecewise fashion to convert a composite uniform B-spline surface to its equivalent explicit polynomial form. Each explicit polynomial patch will be parametrised locally over $[0,1] * [0,1]$.

4.5.3 B-spline (non-uniform) to explicit polynomial: curves

A method for converting the more general B-spline curve to its equivalent explicit polynomial representation is to use the knot insertion algorithm (4.4.7) to convert it first to piecewise Bézier form. This piecewise Bézier curve can then be transformed to explicit polynomial form using the matrix transformation method described in (4.3.1). The resulting explicit polynomial segments will be parametrised over $[0,1]$.

A more direct approach is to carry out the following three steps.

Step 1: Determine the segment breakpoints from the given knot set. Assume this gives us breakpoints $\eta_0, \ldots, \eta_r$ say.

Step 2: Evaluate all derivatives (i.e from 0 to n for a curve of degree n) of the B-spline curve at each of the breakpoints $\eta_0, \ldots, \eta_{r-1}$.

Step 3: For each of the r curve segments, substitute the appropriate derivative values into a Taylor expansion centered about the corresponding parameter breakpoint.

If the original B-spline curve is represented by

$$\mathbf{P}(u) = \sum_{i=0}^{p} \mathbf{w}_i N_{i,n+1}(u)$$

on the knot set $\{u_j\}_{j=0}^{p+n+1}$, the breakpoint sequence is obtained simply by noting those distinct knots occurring in the sequence $\{u_j\}_{j=n}^{p+1}$. The Cox de-Boor algorithm can be used to evaluate the derivatives (see [5]). An efficient method for computing all the derivatives of a B-spline curve at a point is provided in [3].

If we concentrate on one segment of the original B-spline curve, call it $\mathbf{Q}(u)$, where $c \leq u \leq d$ say, Step 2 will result in $n + 1$ coefficients $\mathbf{f}_0, \ldots, \mathbf{f}_n$ say, such that the Taylor expansion about $u = c$ is

$$\mathbf{Q}(u) = \sum_{i=0}^{n} \mathbf{f}_i (u-c)^i / i!$$

$$= \sum_{j=0}^{n} \left(\sum_{k=0}^{n-j} \mathbf{f}_{k+j}(-c)^k / j!k! \right) u^j \quad \text{for } c \le u \le d. \tag{4.25}$$

Note that the use of this conversion method will generate the explicit polynomial segments over the original non-uniform knot set.

4.5.4   B-spline (non-uniform) to explicit polynomial: surfaces

As in the curve case (4.5.3) we can convert a composite B-spline surface on a non-uniform knot set to explicit polynomial form by first using the knot insertion algorithm for surfaces (4.4.9) to convert to Bézier form, and then applying the matrix method (4.3.2) to convert the Bézier patches to polynomial form. Each patch of the resulting polynomial surface will be parametrised locally over $[0,1]*[0,1]$.

We can also generalise the Taylor expansion conversion method to the surface case. If the B-spline surface of degree m by n is represented by

$$\mathbf{P}(u,v) = \sum_{i=0}^{p} \sum_{j=0}^{q} \mathbf{w}_{ij} N_{i,m+1}(u) N_{j,n+1}(v)$$

on the knot set $\{u_i\}_{i=0}^{p+m+1} * \{v_j\}_{j=0}^{q+n+1}$, the individual breakpoint sequences in u and v are determined by noting the distinct knots in the sequences $\{u_i\}_{i=m}^{p+1}$ and $\{v_j\}_{j=n}^{q+1}$ respectively. Assuming that this gives us breakpoints $(\eta_0, \ldots, \eta_r)$ and $(\zeta_0, \ldots, \zeta_s)$ say in u and v (so that there are in total rs patches), we can perform the conversion in three steps.

Step 1: apply the curve algorithm (4.5.3) in the v direction to the rows of control points

$$\mathbf{P}_i(v) = \sum_{j=0}^{q} \mathbf{w}_{ij} N_{j,n+1}(v) \quad \text{for } i = 0, \ldots, p,$$

using the breakpoints $\zeta_0, \ldots, \zeta_{s-1}$. For each of these segment breakpoints, in each row, this process will generate $n+1$ coefficients corresponding to the point, first, second, . . . , n^{th} derivative of $\mathbf{P}_i(v)$. Let this set of coefficients be denoted $\mathbf{b}_{il_1}^{\alpha_1}$ where $\alpha_1 = 0, \ldots, n$ is the index for the derivatives in v and $l_1 = 1, \ldots, s$ is the index for the patches in the v direction.

Step 2: apply the curve algorithm (4.5.3) in the u direction to each subsequent column of coefficients resulting from Step 1 using the breakpoints $\eta_0, \ldots \eta_{r-1}$:

$$\mathbf{P}_j(u) = \sum_{i=0}^{p} \mathbf{b}_{i l_1}^{\alpha_1} N_{i,m+1}(u) \quad \text{for } j = 1, \ldots s.$$

For each one of the $(n+1)s$ columns of coefficients and for each segment breakpoint in u this process will generate $m+1$ coefficients corresponding to the point, first, second, $\ldots$, m^{th} derivative of $\mathbf{P}_j(u)$. Let the set of coefficients so obtained be denoted $\mathbf{c}_{l_1 l_2}^{\alpha_1 \alpha_2}$ where $\alpha_2 = 0, \ldots, m$ is the index for the derivatives in u and $l_2 = 1, \ldots, r$ is the index for the patches in the u direction.

Step 3: for each one of the rs surface patches, substitute the corresponding derivative coefficients $\mathbf{c}_{l_1 l_2}^{\alpha_1 \alpha_2}$ into a 2D Taylor expansion centered about the point $(\eta_{l_2}, \zeta_{l_1})$:

$$\mathbf{Q}_{l_2 l_1}(u, v) = \sum_{i=0}^{m} \sum_{j=0}^{n} \mathbf{c}_{l_1 l_2}^{ij} (u - \eta_{l_2})^i (v - \zeta_{l_1})^j / i! j! \tag{4.26}$$

for $(u, v) \in [\eta_{l_2}, \eta_{l_2+1}] * [\zeta_{l_1}, \zeta_{l_1+1}]$. Equation (4.26) can be expressed in terms of the standard explicit polynomial basis functions, $\{u^i v^j\}_{i,j=0}^{m,n}$, by separating out the double sum for $\mathbf{Q}$,

$$\mathbf{Q}_{l_1 l_2}(u, v) = \sum_{i=0}^{m} \left(\sum_{j=0}^{n} \mathbf{c}_{l_1 l_2}^{ij} (v - \zeta_{l_1})^j / j! \right) (u - \eta_{l_2})^i / i! .$$

We now apply the corresponding curve result (4.25) to the inner sum and then to the subsequent outer sum.

4.5.5 Explicit polynomial to B-spline: curves

If $\mathbf{P}(u)$ is a single segment curve of degree n in explicit polynomial form

$$\mathbf{P}(u) = \sum_{i=0}^{n} \mathbf{a}_i u^i \quad \text{for } u \in [0, 1],$$

and the corresponding B-spline representation is

$$\mathbf{P}(u) = \sum_{i=0}^{n} \mathbf{w}_i N_{i,n+1}(u) \quad \text{for } u \in [0, 1],$$

then from (4.21) we have

$$(\mathbf{w}_0 \quad \dots \quad \mathbf{w}_n)^T = \left[U_n\right]^{-1} (\mathbf{a}_0 \quad \dots \quad \mathbf{a}_n)^T.$$

The cubic case gives the result

$$\begin{pmatrix} \mathbf{w}_0 \\ \mathbf{w}_1 \\ \mathbf{w}_2 \\ \mathbf{w}_3 \end{pmatrix} = \frac{1}{3} \begin{pmatrix} 3 & -3 & 2 & 0 \\ 3 & 0 & -1 & 0 \\ 3 & 3 & 2 & 0 \\ 3 & 6 & 11 & 18 \end{pmatrix} \begin{pmatrix} \mathbf{a}_0 \\ \mathbf{a}_1 \\ \mathbf{a}_2 \\ \mathbf{a}_3 \end{pmatrix}.$$

To convert a composite curve to B-spline form presents some problems as the explicit polynomial representation contains no explicit information regarding the level of cross segment continuity (as was true in the Bézier case). In the special case where the composite curve is parametrised uniformly, increasing by one across each segment, and the continuity level is known to be up to order $n - 1$, only one knot need be assigned at each breakpoint and conversion to a uniform B-spline is appropriate using the above matrix method. The amount of computation can be reduced by noting that $n - 1$ B-spline control points will be common to each pair of neighbouring segments.

In the more general case where the continuity level across the segment boundaries is not known, one possibility is to convert to Bézier form (4.3.3), and interpret it as a B-spline with the appropriate knot multiplicities. No continuity information above C^0 will be transmitted using this method. If the continuity level can be determined, perhaps by evaluating left and right derivatives up to order $n - 1$ at the segment boundaries and checking for agreement within a set tolerance, then a minimal knot set for the composite curve can be constructed. In particular, if we begin with the corresponding Bézier knot set, for each left and right derivative that agrees at a particular breakpoint we can reduce the multiplicity of the corresponding knot by one. Having found the minimal knot set for the curve the B-spline control points can be determined using the following de Boor-Fix algorithm [6].

4.5.6   de Boor-Fix algorithm

If

$$\mathbf{P}(u) = \sum_{i=0}^{p} \mathbf{w}_i N_{i,n+1}(u) \tag{4.27}$$

is the required B-spline representation on the knot set $(u_0, \dots, u_{p+n+1})$, and we define the linear operator λ_j by its effect on a function f as

$$\lambda_j(f) = \frac{1}{n+1} \sum_{r=0}^{n} (-1)^{n-r} \Psi_j^{(r)}(a_j) f^{(n-r)}(a_j), \tag{4.28}$$

(here the superscript in brackets denotes the derivative of that order) where a_j is any point in the open interval (u_j, u_{j+n+1}) and

$$\Psi_j(y) = \prod_{r=1}^{n} (y - t_{j+r}),$$

then (see [6] page 116 for proof)

$$\lambda_j(N_{i,n+1}(u)) = \delta_{ij} = \begin{cases} 1 & i = j \\ 0 & i \neq j \end{cases}.$$

Applying the operators λ_j to (4.27) we obtain

$$\mathbf{w}_j = \lambda_j(\mathbf{P}(u)) \quad \text{for } j = 0, \ldots, p.$$

Hence if we apply the λ_j for $j = 0, \ldots, p$ to $\mathbf{P}(u)$ in its explicit polynomial form we will generate the corresponding B-spline control points.

4.5.7 Explicit polynomial to B-spline: surfaces

For a single surface patch of degree m by n we have from equation (4.24)

$$\left[\mathbf{w}_{ij}\right] = \left[U_m\right]^{-1} \left[\mathbf{a}_{ij}\right] \left[U_n\right]^{-1T}. \tag{4.29}$$

The problems mentioned in (4.5.5) in attempting to convert a composite explicit polynomial curve to B-spline form apply equally well in the surface case. We can as in the curve case convert each patch to Bézier form and interpret the resulting surface as a B-spline with the appropriate knot multiplicity set in u and v. Again only C^0 continuity will be conveyed by this representation. If the surface is known to be continuous up to order $m - 1$ in u and $n - 1$ in v and the parametrisation is uniform integer in both directions we can convert to the uniform B-spline form using (4.29). If the surface is known to be C^1 in both directions and the patches are parametrised locally over $[0, 1] * [0, 1]$ it is possible to convert each one to Bézier form as described in (4.3.4) and then to apply the C^1 Bézier to B-spline algorithm in (4.4.12).

Conversion in the more general case requires the determination of the minimal knot set for u and v. This could be obtained by evaluating cross boundary derivatives at the breakpoints along the u and v directions and checking for equality within a set tolerance. The minimum cross boundary continuity between two patches along a particular row or column will determine the appropriate knot multiplicity to assign at the corresponding breakpoint in u or v. Having obtained a knot set for u and a knot set for v the B-spline control points can be obtained by applying the de Boor-Fix curve algorithm (4.5.6). In particular, if the B-spline surface is represented by

$$\mathbf{P}(u,v) = \sum_{i=0}^{p} \sum_{j=0}^{q} \mathbf{w}_{ij} N_{i,m+1}(u) N_{j,n+1}(v),$$

and we have corresponding linear operators $(\lambda_\alpha)_{\alpha=0}^{p}$, $(\eta_\beta)_{\beta=0}^{q}$ with

$$\lambda_\alpha\Big(N_{i,m+1}(u)\Big) = \delta_{i\alpha} \text{ and } \mu_\beta\Big(N_{j,n+1}(v)\Big) = \delta_{j\beta},$$

then

$$\mathbf{w}_{\alpha\beta} = \sum_{i=0}^{p} \sum_{j=0}^{q} \mathbf{w}_{ij} \, \lambda_\alpha\Big(N_{i,m+1}(u)\Big) \mu_\beta\Big(N_{j,n+1}(v)\Big).$$

Hence the B-spline control points can be found by applying the operators $\lambda_\alpha, \mu_\beta$ to $\mathbf{P}(u,v)$ in its explicit polynomial form and can be achieved in two steps:

Step 1: Consider each row of polynomial coefficients in the matrix of patches as a composite polynomial curve in v. Apply the curve algorithm in (4.5.6) to each of these rows using the operators η_β for $\beta = 0,\ldots,q$.

Step 2: Consider each subsequent column of coefficients resulting from Step 1 as a composite polynomial curve in u and apply the curve algorithm in (4.5.6) to each of these columns using the operators λ_α for $\alpha = 0,\ldots,p$. The points resulting from this second step are the required B-spline control points.

Appendix 1 to Chapter 4

We wish to derive the relationship (4.13) between the two sets of B-spline basis functions $\{N_{i,k}(u)\}$ and $\{\hat{N}_{i,k}(u)\}$. Recall that the second set differs from the first in that it has had one extra knot inserted at $\hat{u} = u_{l+1}$.

The following definition of the B-spline basis function $N_{i,k}(u)$ is taken from [5]:

Let

$$g_k(s; u) = (s - u)_+^{k-1} = \begin{cases} (s - u)^{k-1} & s \geq u \\ 0 & s < u \end{cases},$$

then the *standardised* B-spline basis function $M_{i,k}(u)$ is defined to be the kth divided difference of $g_k(s; u)$ in s on $u_i, \ldots, u_{i+k}$ for fixed u:

$$M_{i,k}(u) = g_k[u_i, \ldots, u_{i+k}; u]$$

$$= \frac{g_k[u_{i+1}, \ldots, u_{i+k}; u] - g_k[u_i, \ldots, u_{i+k-1}; u]}{u_{i+k} - u_i}. \qquad (4.31)$$

The *normalised* B-spline basis function $N_{i,k}(u)$ is defined as

$$N_{i,k}(u) = (u_{i+k} - u_i)M_{i,k}(u). \qquad (4.32)$$

Using (4.31) we have the following result,

$$(\hat{u} - u_i)g_k[u_i, \ldots, u_{i+k-1}, \hat{u}; u] + (u_{i+k} - \hat{u})g_k[\hat{u}, u_{i+1}, \ldots, u_{i+k}; u]$$

$$-(u_{i+k} - u_i)g_k[u_i, \ldots, u_{i+k}; u] = 0 \quad \text{for} \quad i = l - k + 1, \ldots, l.$$

Hence

$$(u_{i+k} - u_i)M_{i,k}(u) = (\hat{u} - u_i)\hat{M}_{i,k}(u) + (u_{i+k} - \hat{u})\hat{M}_{i+1,k}(u),$$

or using (4.32),

$$N_{i,k}(u) = \frac{\hat{u} - \hat{u}_i}{\hat{u}_{i+k} - \hat{u}_i}\hat{N}_{i,k}(u) + \frac{\hat{u}_{i+k+1} - \hat{u}}{\hat{u}_{i+k+1} - \hat{u}_{i+1}}\hat{N}_{i+1,k}(u)$$

for $i = l - k + 1, \ldots, l$ as was to be shown.

Note that if u_{l+1} occurs with multiplicity s then $\hat{u} = \hat{u}_{l+1} = \ldots = \hat{u}_{l+s+1}$ and so $\hat{N}_{i,k}(u) = N_{i,k}(u)$ for $i \leq l - k + s$ and $\hat{N}_{i,k}(u) = N_{i+1,k}(u)$ for $i \geq l + 1$. Figure 25 illustrates the effect of inserting of $\hat{u} = 1$ into the cubic B-spline knot set $(0, 0, 0, 0, 1, 1, 2, 2, 3, 3, 3, 3)$.

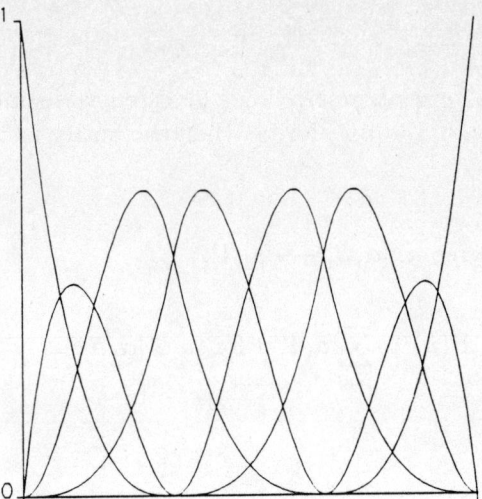

Figure 25a. Cubic B-spline basis functions on knot set $(0,0,0,0,1,1,2,2,3,3,3,3)$

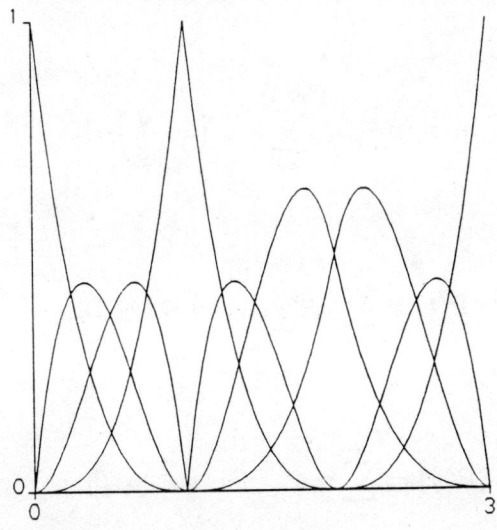

Figure 25b. Cubic B-spline basis functions on knot set $(0,0,0,0,1,1,1,2,2,3,3,3,3)$

Appendix 2 to Chapter 4

We list here the formulae for reparametrising a given curve segment or surface patch to and from the [0,1] and $[0,1] * [0,1]$ forms when the entity is expressed in explicit polynomial form.

(a) Curve segment of degree n, $[\alpha, \beta] \longrightarrow [0,1]$:

$$\mathbf{P}(u) = \sum_{i=0}^{n} \mathbf{a}_i u^i \quad \text{for } u \in [\alpha, \beta].$$

For $t = (u - \alpha)/(\beta - \alpha)$

$$\mathbf{P}(t) = \sum_{i=0}^{n} \mathbf{b}_i t^i \quad \text{for } t \in [0, 1],$$

where

$$\mathbf{b}_i = (\beta - \alpha)^i \sum_{j=0}^{n-i} \mathbf{a}_{i+j} \binom{i+j}{i} \alpha^j.$$

(b) $[0,1] \longrightarrow [\alpha, \beta]$:

$$\mathbf{P}(t) = \sum_{i=0}^{n} \mathbf{a}_i t^i \quad \text{for } 0 \le t \le 1.$$

For $u = t(\beta - \alpha) + \alpha$

$$\mathbf{P}(u) = \sum_{i=0}^{n} \mathbf{b}_i u^i \quad \text{for } u \in [\alpha, \beta],$$

where

$$\mathbf{b}_i = \sum_{j=0}^{n-i} \frac{\mathbf{a}_{i+j}}{(\beta - \alpha)^{i+j}} \binom{i+j}{i} (-\alpha)^j.$$

(c) Surface patch of degree m by n, $[\alpha, \beta] * [\eta, \zeta] \longrightarrow [0,1] * [0,1]$:

$$\mathbf{P}(u, v) = \sum_{i=0}^{m} \sum_{j=0}^{n} \mathbf{a}_{ij} u^i v^j \quad \text{for } (u, v) \in [\alpha, \beta] * [\eta, \zeta].$$

Step 1: apply (a) for $[\eta, \zeta] \longrightarrow [0,1]$ to the rows of coefficients

$$\mathbf{P}_i(v) = \sum_{j=0}^{n} \mathbf{a}_{ij} v^j \quad \text{for } i = 0, \ldots, m.$$

Step 2: apply (a) for $[\alpha, \beta] \longrightarrow [0,1]$ to the columns of coefficients

$$\mathbf{P}_j(u) = \sum_{i=0}^{m} \mathbf{b}_{ij} u^i \quad \text{for } j = 0, \ldots, n,$$

where the $\mathbf{b}_{ij}$ are the coefficients resulting from Step 1.

(d) $[0,1] * [0,1] \longrightarrow [\alpha, \beta] * [\eta, \zeta]$:

$$\mathbf{P}(s,t) = \sum_{i=0}^{m} \sum_{j=0}^{n} \mathbf{a}_{ij} s^i t^j \quad \text{for } (s,t) \in [0,1] * [0,1].$$

Step 1: apply (b) for $[0,1] \longrightarrow [\eta, \zeta]$ to the rows of coefficients

$$\mathbf{P}_i(t) = \sum_{j=0}^{n} \mathbf{a}_{ij} t^j \quad \text{for } i = 0, \ldots, m.$$

Step 2: apply (b) for $[0,1] \longrightarrow [\alpha, \beta]$ to the columns of coefficients

$$\mathbf{P}_j(s) = \sum_{i=0}^{m} \mathbf{b}_{ij} s^i \quad \text{for } j = 0, \ldots, n,$$

where the $\mathbf{b}_{ij}$ are the coefficients resulting from Step 1.

5   DEGREE REDUCTION APPROXIMATIONS

Principal authors: M.A. Lachance, P.A. Sherar, R.J. Goult

5.0   Introduction

The majority of CAD systems provide some form of parametric curve and surface representation but the precise form of this representation varies considerably.
Some systems use simple parametric polynomial curves and surfaces, others provide some form of rational parametric representation. Even amongst those systems
using parametric polynomials they differ in the maximum degree allowed and the
form of representation, which may be explicit polynomial base, B-spline or Bézier.
Provided the degrees are the same, exact conversions between these different representations are possible as detailed in Chapter 4 but approximation problems arise
when transferring data from a rational or other non-polynomial based system to a
polynomial system. The generalised Chebyshev technique described in this chapter addresses the degree reduction problem, the orthogonal polynomial methods
described in Chapter 6 can be used for all other types of approximation problem.

The two techniques have some important features in common as well as some significant differences. Both methods produce a polynomial approximation to a given
curve or surface which in some sense is as close as possible to the original both in
geometric shape and in parametrisation. The use of the constrained polynomials in
both techniques enables end conditions on parametric curves and edge conditions on
parametric surfaces to be preserved in the approximation. This is the most important feature of the methods, since it enables continuity conditions to be maintained
when approximating separate segments of spline curves or patches of a piecewise
defined surface. An easily computed measure of the error in the approximation is
available from both methods; this can be used as part of an automatic strategy to
produce an approximation which satisfies a user-defined tolerance threshold. The
generalised Chebyshev economisation technique can be regarded as a destructive
method in which, in order to produce a low degree approximation to a parametric
polynomial curve or surface, successive approximations are produced in which the
degree is reduced by one at each step. The orthogonal polynomial method is essentially a constructive technique in which the low degree polynomial approximations
can be obtained first, although in practice it is usual to go straight to the target
degree approximation.

In this chapter we describe the method of generalised Chebyshev economisation.
The use of the constrained Chebyshev polynomials developed for this purpose pro-

vides an effective method of degree reduction with pre-determined levels of interpolation at the ends of curve segments or at the corners of surface patches. The properties of these polynomials ensure that the approximation obtained is optimal under the uniform norm, and provide a readily computed error estimate. Practical experience has confirmed the theoretical advantages of this approach.

5.1   Constrained Chebyshev polynomials

The classical Chebyshev polynomials can be defined, for each positive integer m, by the equation

$$T_m(cos\,\theta) = cos\,(m\theta) \quad \text{for} \quad 0 \leq \theta \leq \pi. \tag{5.1}$$

The importance of these polynomials in approximation theory was established by Chebyshev when he proved that the polynomial $g(t)$ of degree at most $m-1$ which minimises

$$E_m = \max_{t \in [-1,1]} |\,t^m - g(t)\,| \tag{5.2}$$

is $t^m - 2^{1-m}T_m(t)$. The corresponding value of E_m is 2^{1-m} and this is attained precisely $m+1$ times in the interval [-1,1] by virtue of the equi-oscillatory property of $T_m(t)$ (see Figure 26). In the classical Chebyshev economisation method the degree of

$$f(t) = \sum_{i=0}^{m} a_i t^i$$

is reduced to at most $m-1$ by defining

$$h(t) = f(t) - a_m 2^{1-m}T_m(t); \tag{5.3}$$

then $h(t)$ is the reduced degree polynomial closest to $f(t)$ in the sense of the uniform norm. The properties of $T_m(t)$ ensure that the maximum absolute deviation is $2^{1-m}\,|\,a_m\,|$ and that there is zero error at precisely m points in the interval $[-1,1]$. Since $T_m(t)$ is non-zero at $t = -1$ and $t = 1$, $f(t)$ and $h(t)$ will not coincide at the end points; for this reason classical Chebyshev economisation cannot maintain continuity when applied to parametric spline curves.

The desired continuity conditions can be attained by replacing the Chebyshev polynomials by the constrained Chebyshev polynomials first introduced in [15]. For convenience of application to parametric curves and surfaces the interval $[0, 1]$ will

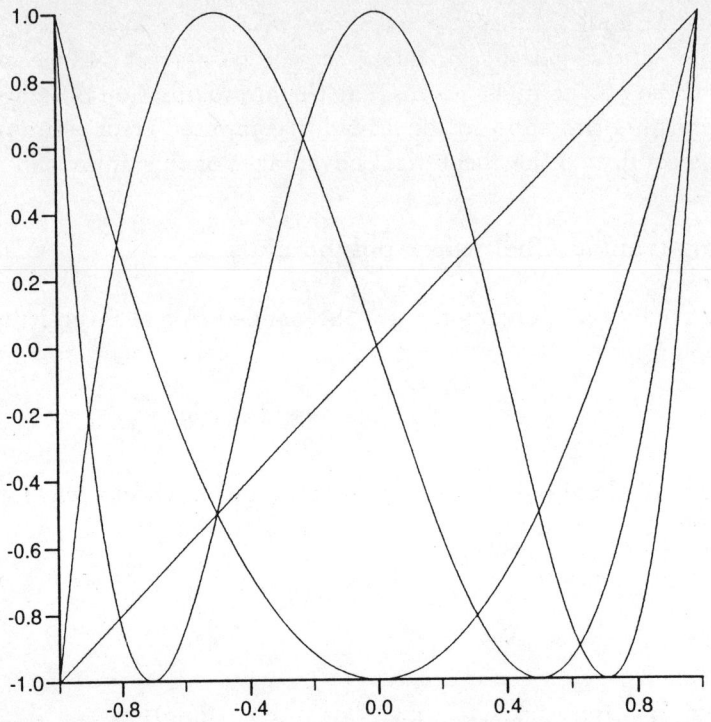

Figure 26. Classical Chebyshev polynomials, degrees 1 to 4

be used in place of the standard $[-1, 1]$ interval. By analogy with (5.2) we seek the polynomial $g(t)$ of degree at most $m - 2k - 1$ which minimises

$$E_m^k = \max_{t \in [0,1]} | \, t^k (1 - t)^k (t^{m-2k} - g(t)) \, | . \tag{5.4}$$

The normalised constrained Chebyshev polynomial for $m > 2k$ is then defined as

$$C_m^k(t) = t^k (1 - t)^k (t^{m-2k} - g(t)). \tag{5.5}$$

$C_m^k(t)$ is a monic polynomial of degree m with k-fold zeros at $t = 0$ and at $t = 1$ whose uniform norm on $[0, 1]$ is as small as possible. $C_m^k(t)$ also has the equi-oscillatory properties of the standard Chebyshev polynomials and its modulus attains the bound E_m^k precisely $m - 2k + 1$ times in the interval $[0, 1]$. For values of k greater than 1, $C_m^k(t)$ cannot be obtained explicitly but can be found numerically via a modified Remes algorithm which exploits the equi-oscillatory behaviour. Figures 27, 28 and 29 show $C_m^1(t)$, $C_m^2(t)$ and $C_m^3(t)$ for various values of m.

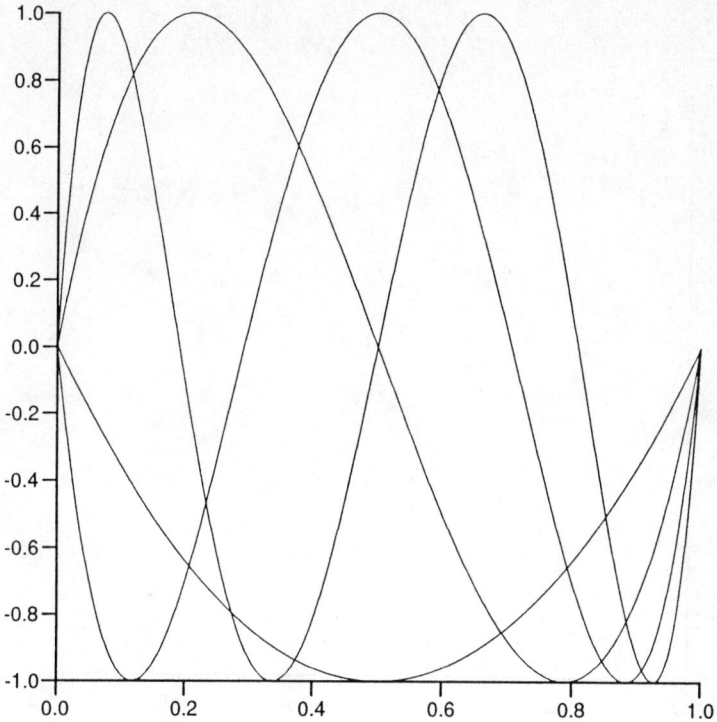

Figure 27. Point constrained Chebyshev polynomials, degrees 2 to 5

5.2 Parametric curve approximation using constrained Chebyshev polynomials

A parametric polynomial curve can be defined as

$$\mathbf{r}(t) = \sum_{i=0}^{m} \mathbf{a}_i t^i \quad \text{for} \quad 0 \le t \le 1, \tag{5.6}$$

where the $\mathbf{a}_i$ are vector coefficients. By an obvious generalisation of Chebyshev economisation, a suitable approximation of degree at most $m - 1$ is

$$\mathbf{s}(t) = \mathbf{r}(t) - \mathbf{a}_m C_m^k(t). \tag{5.7}$$

The properties of $C_m^k(t)$ ensure that $\mathbf{r}(t)$ and $\mathbf{s}(t)$ coincide at the points where $t = 0$ and $t = 1$ and that they have $k - 1$ coincident derivatives at these points. $\mathbf{s}(t)$ is the best such approximation in the sense of minimising, for $t \in [0, 1], |\mathbf{r}(t) - \mathbf{s}(t)|$. The maximum value of the Euclidean norm $|\mathbf{r}(t) - \mathbf{s}(t)|$ is $E_m^k |\mathbf{a}_m|$.

Using this technique $\mathbf{s}(t)$ approximates both the geometry and the parametrisation of $\mathbf{r}(t)$ although better geometric approximations might be obtained by ignoring the

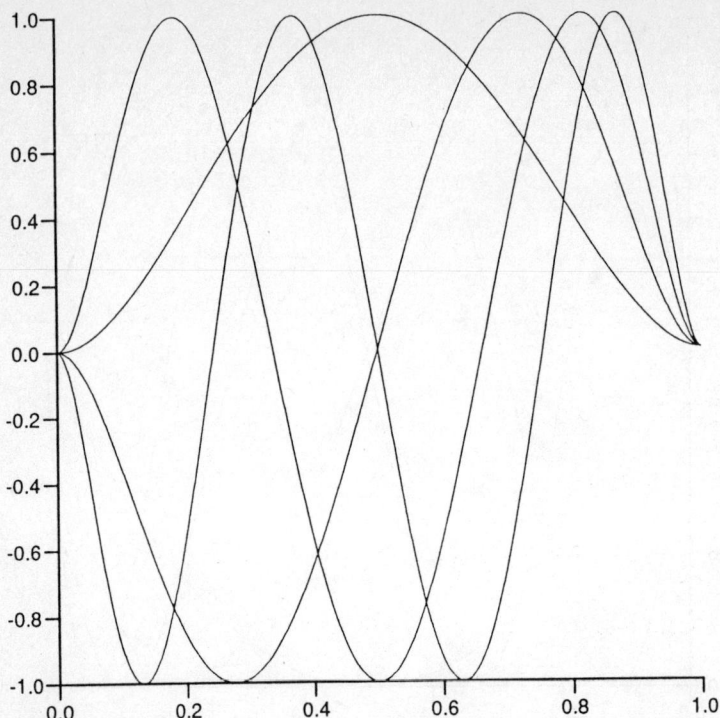

Figure 28. First derivative constrained Chebyshev polynomials, degrees 3 to 6

parametrisation. The $m - 2k$ roots of $C_m^k(t)$ between 0 and 1 ensure that $\mathbf{s}(t)$ will interpolate $\mathbf{r}(t)$ at $m - 2k$ points within this range. For most CAD applications a degree reduction of more than one is required. This can be obtained by successive applications of the method described above. If, for example, a reduction of n degrees is required, the first approximation, $\mathbf{s}_1(t)$ of degree at most $m - 1$ is given directly by (5.7). If $\mathbf{a}_{m-1}^*$ is the coefficient of t^{m-1} in $\mathbf{s}_1(t)$ then we define

$$\mathbf{s}_2(t) = \mathbf{s}_1(t) - \mathbf{a}_{m-1}^* C_{m-1}^k(t)$$

and, for $1 \leq i \leq n - 1$, the iterative definition is

$$\mathbf{s}_{i+1}(t) = \mathbf{s}_i(t) - \mathbf{a}_{m-i}^* C_{m-i}^k(t), \tag{5.8}$$

where $\mathbf{a}_{m-i}^*$ is the leading vector coefficient in $\mathbf{s}_i(t)$. The final approximation is $\mathbf{s}(t) = \mathbf{s}_n(t)$ with an error bound given by

$$\mid \mathbf{r}(t) - \mathbf{s}(t) \mid \leq \mid \mathbf{a}_m \mid E_m^k + \mid \mathbf{a}_{m-1}^* \mid E_{m-1}^k + \ldots + \mid \mathbf{a}_{m-n+1}^* \mid E_{m-n+1}^k. \tag{5.9}$$

Unlike the simple case of a single degree reduction step the error bound given by (5.9) is not generally attained and $\mathbf{s}(t)$ will not necessarily interpolate $\mathbf{r}(t)$ at

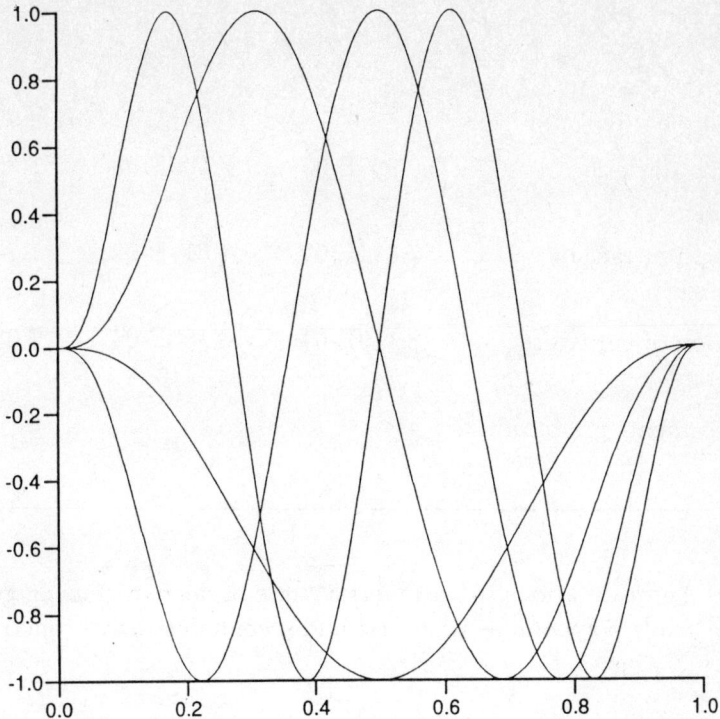

Figure 29. Second derivative constrained Chebyshev polynomials, degrees 4 to 7

any points between $t = 0$ and $t = 1$. In practice (5.9) has been found to be sufficiently close to the true approximation error to be useful in controlling a semi-automatic approximation process with predefined target degree, continuity level and error bound limits. If after applying the method defined by (5.8) the predicted error bounds are unacceptably large then the range must be subdivided and by applying the above technique to separate parts of the curve $\mathbf{r}(t)$ is approximated by a piecewise parametric polynomial of the target degree. The properties of $C_m^k(t)$ at the end points ensure this is done with C^{k-1} continuity between the segments. A simple strategy to obtain an approximation with predefined accuracy is to bisect a segment at the parametric mid-point whenever (5.9) indicates that the error limits have been exceeded.

Table 3 shows some sample results obtained with this technique. For this example the Chebyshev polynomials $C_m^3(t)$ were used so that both first and second derivative continuity is maintained with the original curve at the endpoints. This is also the level of cross segment continuity in the economisations. The errors in the tangent vector direction have been monitored and it is interesting that, in spite of the highly oscillatory nature of the constrained Chebyshev polynomials, $\mathbf{s}(t)$ gives a smooth approximation to the original curve.

target degree	9	9	9
selected accuracy [mm]	0.1	0.01	0.001
max. absolute deviation (inside) [mm]	9.997E-03	9.505E-03	1.801E-04
max. deviation of tangent (inside) [degree]	7.611E-02	7.611E-02	1.740E-03
max. deviation of curvature (inside) [1/mm]	8.113E-04	3.168E-04	7.331E-06
number of segments produced	4	5	6

Table 3. Approximations to parametric curve of degree 13 using generalised Chebyshev polynomials with first and second derivative constraints (k=3)

5.3 Surface approximation using constrained Chebyshev polynomials

The technique described in section 5.2 has been generalised in order to approximate bi-parametric polynomial surfaces of the tensor product type. In the application to patched parametric surfaces an important property to be preserved is consistent approximation to the common boundary curves when approximating adjacent surface patches. Conventionally a bi-parametric polynomial surface of degree m by n in the parameters u and v respectively, is defined as

$$\mathbf{P}(u,v) = \sum_{i=0}^{m} \sum_{j=0}^{n} \mathbf{a}_{ij} u^i v^j \quad \text{for} \quad [u,v] \in [0,1] * [0,1], \qquad (5.10)$$

where the $\mathbf{a}_{ij}$ are the vector coefficients. Two alternative representations which separate the two variables u and v are

$$\mathbf{P}(u,v) = \sum_{i=0}^{m} \mathbf{P}_i(v) u^i,$$

where

$$\mathbf{P}_i(v) = \sum_{j=0}^{n} \mathbf{a}_{ij} v^j,$$

and

$$\mathbf{P}(u,v) = \sum_{j=0}^{n} \mathbf{P}^j(u) v^j,$$

where

$$\mathbf{P}^j(u) = \sum_{i=0}^{m} \mathbf{a}_{ij} u^i.$$

For any pair of positive integers (l, k) with $2l \leq m$ and $2k \leq n$ the constrained Chebyshev economisation surface is defined by the equation

$$\mathbf{Q}(u,v) = \mathbf{P}(u,v) - \mathbf{P}_m(v)C_m^k(u) - \mathbf{P}^n(v)C_n^l(v) + \mathbf{a}_{mn}C_m^k(u)C_n^l(v). \qquad (5.11)$$

It can be verified that $\mathbf{Q}(u,v)$ is a parametric surface of degree at most $m - 1$ in u by $n - 1$ in v. The properties of the constrained Chebyshev polynomials ensure that at the patch corners $\mathbf{Q}(u,v)$ matches $\mathbf{P}(u,v)$ in value and in derivatives according to the selected values of k and l. Considering the behaviour of $\mathbf{Q}(u,v)$ along a boundary we obtain, along $v = 0$ for example,

$$\mathbf{Q}(u,0) = \mathbf{P}(u,0) - \mathbf{P}_m(0)C_m^k(u)$$

$$= \sum_{i=0}^{m} \mathbf{a}_{i0} u^i - \mathbf{a}_{m0} C_m^k(u). \qquad (5.12)$$

Equation (5.12) is identical to the economisation of the boundary curve $\mathbf{P}(u,0)$ as described in section 5.2. Hence any two adjacent surfaces with matching boundary curves will, after degree reduction, retain matching boundary curves. It can also be shown [16] that if we begin with a pair of adjacent surfaces that possess a certain level of cross boundary continuity, then this continuity can be preserved in the economisations. In particular, if k and l are both greater than zero, simple continuity across the two approximations is assured and if they are both greater than one, cross derivative continuity holds, and so on.

If degree reduction is required in only one of the two parameters, as would be the case in seeking an economisation of a ruled surface, then we can replace equation (5.12) by either

$$\mathbf{Q}(u,v) = \mathbf{P}(u,v) - \mathbf{P}_m(v)C_m^k(u) \tag{5.13}$$

for a one degree reduction in the u direction, or

$$\mathbf{Q}(u,v) = \mathbf{P}(u,v) - \mathbf{P}^n(u)C_n^l(v) \tag{5.14}$$

for a one degree reduction in v. Applying a one degree reduction in u (or v) using (5.13) and then a one degree reduction in v (or u) using (5.14) results in the same approximating surface as that defined by (5.11).

A bound on the error in the approximation is given by

$$|\mathbf{Q} - \mathbf{P}| \le \max_{v \in [0,1]} |\mathbf{P}_m(v)| E_m^k + \max_{u \in [0,1]} |\mathbf{P}^n(u)| E_n^l + |\mathbf{a}_{mn}| E_m^k E_n^l. \tag{5.15}$$

Unfortunately, due to the fact that the above bound depends on u and v, this expression is not so easy to deal with computationally. To obtain a form that is we first express $\mathbf{P}(u,v)$ with respect to the Bézier-Bernstein basis:

$$\mathbf{P}(u,v) = \sum_{i=0}^{m} \sum_{j=0}^{n} \mathbf{v}_{ij} B_{m,i}(u) B_{n,j}(v) \quad \text{for } (u,v) \in [0,1] * [0,1],$$

where

$$B_{m,i}(u) = \binom{m}{i} u^i (1-u)^{m-i} = (-1)^{m-i} \binom{m}{i} u^m + \cdots \quad \text{for} \quad 0 \le i \le m.$$

We recall that $\mathbf{P}_m(v)$ is defined as the coefficient of u^m in $\mathbf{P}(u,v)$ and hence

$$\mathbf{P}_m(v) = \sum_{i=0}^{m} \sum_{j=0}^{n} (-1)^{m-i} \binom{m}{i} \mathbf{v}_{ij} B_{n,j}(v)$$

$$= \sum_{j=0}^{n} \mathbf{w}_j B_{n,j}(v) \quad \text{say,} \tag{5.16}$$

where

$$\mathbf{w}_j = \sum_{i=0}^{m} (-1)^{m-i} \binom{m}{i} \mathbf{v}_{ij}.$$

In a similar fashion we can write

$$\mathbf{P}^n(u) = \sum_{i=0}^{m} \mathbf{w}^i B_{m,i}(u), \tag{5.17}$$

where

$$\mathbf{w}^i = \sum_{j=0}^{n} (-1)^{n-i} \binom{n}{j} \mathbf{v}_{ij}.$$

Using (5.16) and (5.17) in the error bound expression (5.15) and using the fact that the Bézier-Bernstein basis functions sum to one,

$$\sum_{i=0}^{m} B_{m,i}(u) = \sum_{j=0}^{n} B_{n,j}(v) \equiv 1,$$

we obtain the following bound for the difference between $\mathbf{P}$ and $\mathbf{Q}$ (which in turn bounds (5.14)):

$$|\mathbf{Q} - \mathbf{P}| \le \max_{0 \le j \le n} |\mathbf{w}_j| E_m^k + \max_{0 \le i \le m} |\mathbf{w}^i| E_n^l + |\mathbf{a}_{mn}| E_m^k E_n^l. \tag{5.18}$$

Experimental evidence [16] has indicated that the discrepancy between (5.18) and the true uniform error between a given surface and its Chebyshev economisation is not large. Hence it is feasible to use (5.18) as part of a subdivision strategy to obtain an approximating surface of any required degree that lies within a user defined tolerance of the original.

A fairly simple subdivision strategy, which can be used when a single surface approximation fails to meet the specified bound, is to use the curve subdivision algorithm to determine a common subdivision of the u boundary curves $u = 0$ and $u = 1$ and similarly for the v boundary curves . By propagating these across the surface to obtain a topologically rectangular matrix of patches we apply the Chebyshev economisation procedure in a peicewise fashion to each patch. Should, for a particular patch, the bound (5.18) exceed the tolerance, the row and column index of the patch is noted. These patches can be further subdivided by, for example, splitting at the mid-point in the u or v direction in which the patch boundary economisation produces the largest error bound (5.9). In order to maintain a rectangular topology for the final surface the subsequent subdivision of the offending patches must be propagated in the appropriate direction. The economisation however need only be repeated on patches for which (5.18) exceeds the tolerance. All other successful approximations can be stored away immediately or, if need be, after subsequent subdivision. Experimental evidence has indicated that the 'first pass' is often (but by no means always) sufficient, in which case the initial patch structure determined by the economisation of the original boundary curves is the final one.

We illustrate in the following figures the above economisation and subdivision process working on some test data kindly provided by BMW. Figure 30 shows the car with the selected surfaces, a single patch polynomial surface of degree 14 by 14 and one of degree 9 by 7, highlighted. Figure 31 shows the result of economising the first patch to a bi-cubic surface for user defined tolerances of 1.0mm, 0.1mm and 0.01mm. In this example k and l were chosen to be two, so that the approximations preserve tangent continuity with the original surface at the patch corners and so that the economised surface as a whole is tangent continuous. Clearly, as this example shows, reducing the tolerance will in general increase the number of patches produced. Table 4 provides a more detailed analysis of the quality of these approximations.

Finally, we vary the continuity level specified for the bi-quintic economisation of the second patch of degree 9 by 7, for a user defined tolerance of 0.01mm. The sample results are displayed in Table 5. As one might expect, the number of patches generated by the approximation process goes up with increasing order of constraint imposed.

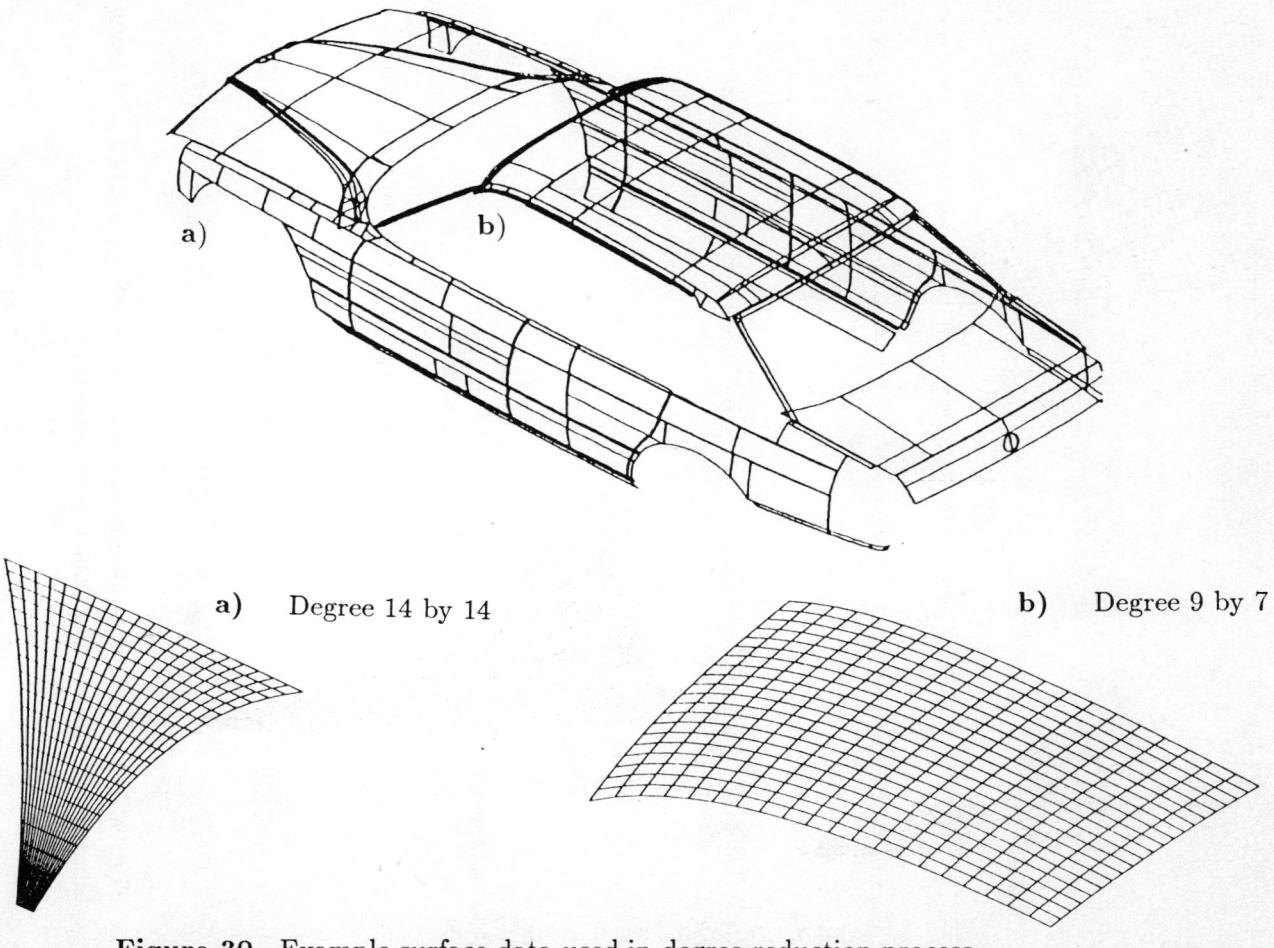

a)   Degree 14 by 14

b)   Degree 9 by 7

Figure 30. Example surface data used in degree reduction process

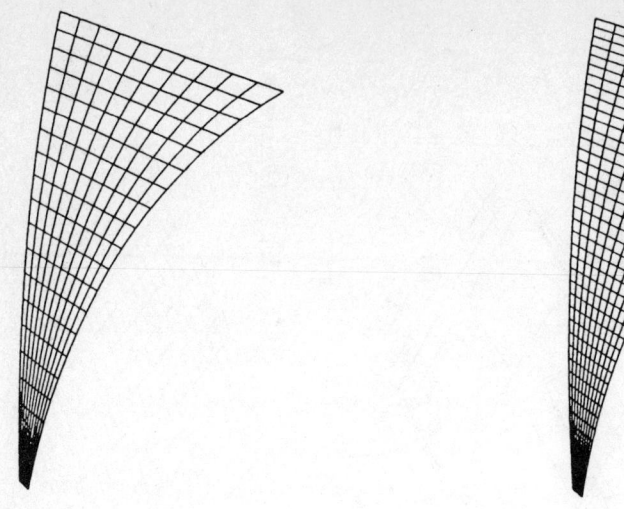
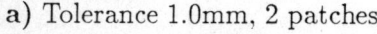
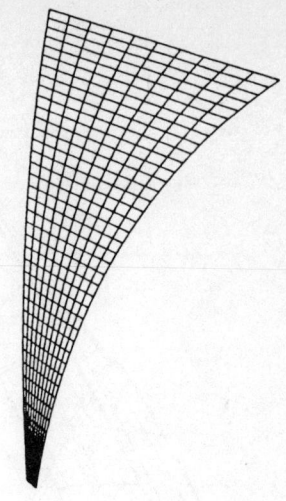

a) Tolerance 1.0mm, 2 patches   b) Tolerance 0.1mm, 4 patches

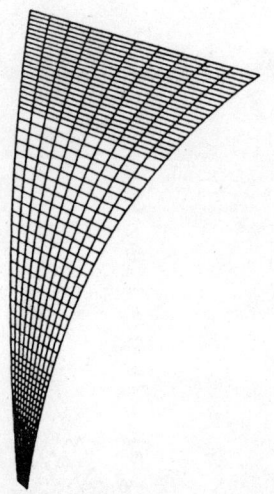

c) Tolerance 0.01mm, 6 patches

Figure 31.   Bi-cubic approximations to patch of degree 14 by 14 using tangent constrained Chebyshev polynomials

target degree	3 by 3	3 by 3	3 by 3
selected accuracy [mm]	1.0	0.1	0.01
selected continuity level	C^1	C^1	C^1
max. absolute deviation (inside) [mm]	0.145	1.362E-02	5.720E-03
max. absolute deviation (u-edge) [mm]	0.159	1.450E-02	6.408E-03
max. absolute deviation (v-edge) [mm]	2.453E-03	2.453E-03	2.453E-03
max. deviation of normal vector (inside) [degree]	7.528E-02	1.513E-02	1.513E-02
max. deviation of curvature u- direction (inside) [1/mm]	2.333E-04	8.828E-05	3.182E-05
max. deviation of curvature u- direction (edge) [1/mm]	4.202E-04	1.417E-04	6.858E-05
max. deviation of curvature v- direction (inside) [1/mm]	1.091E-05	2.222E-06	1.236E-06
max. deviation of curvature v- direction (edge) [1/mm]	2.954E-06	2.954E-06	2.954E-06
number of patches produced	2 (1*2)	4 (1*4)	6 (1*6)

Table 4. Sample results from bi-cubic economisations of degree 14 by 14 surface patch using tangent constrained Chebyshev polynomials

target degree	5 by 5	5 by 5	5 by 5
selected accuracy [mm]	0.01	0.01	0.01
selected continuity level	C^0	C^1	C^2
max. absolute deviation (inside) [mm]	1.663E-03	2.386E-03	1.327E-03
max. absolute deviation (u-edge) [mm]	1.626E-03	1.752E-03	9.673E-03
max. absolute deviation (v-edge) [mm]	7.113E-04	2.402E-03	7.631E-04
max. deviation of normal vector (inside) [degree]	5.806E-03	4.208E-03	1.184E-03
max. deviation of curvature u- direction (inside) [1/mm]	1.810E-06	3.873E-06	5.689E-07
max. deviation of curvature u- direction (edge) [1/mm]	2.331E-05	1.400E-05	6.653E-07
max. deviation of curvature v- direction (inside) [1/mm]	1.679E-06	4.844E-06	5.561E-07
max. deviation of curvature v- direction (edge) [1/mm]	2.600E-05	1.470E-05	5.518E-07
number of patches produced	2 (2*1)	3 (3*1)	8 (4*2)

Table 5. Sample results from bi-quintic economisations of degree 9 by 7 surface patch for varying levels of constraint

6 MORE GENERAL CURVE AND SURFACE APPROXIMATIONS

Principal author: R.J. Goult

6.1 Parametric curve and surface approximation using orthogonal functions

The Chebyshev economisation technique described in chapter 5 is restricted in its application to degree reduction of parametric polynomial curves and surfaces. In practice rational, trigonometric or procedurally defined parametric surfaces are frequently used in CAD systems, and some more general technique of polynomial approximation is needed for data exchange applications. In general the problem is to find a polynomial approximation of specified degree, according to the capabilities of the receiving system, which gives an acceptable fit to the original data. One well known criterion for 'goodness of fit' is the least squares criterion. This can be combined with appropriate orthogonal polynomials, in order to avoid the numerical instability problems commonly found in solving the normal equations associated with the simple least squares method.

In practice the problem is usually rather more complex, in that the approximation is required to satisfy additional constraints. In this chapter modified orthogonal polynomials are developed which can be used for least squares approximation to parametric curves and surfaces with fixed constraints at end points and boundaries. These constraints are necessary in order to ensure the continuity of piecewise parametric curves and patched parametric surfaces. A further benefit of this approach is the provision, at negligible computational cost, of reliable error estimates.

6.2 Orthogonal polynomials and the least squares criterion

We assume $f(t)$ is the function to be approximated over the range $0 \leq t \leq 1$ while $g(t)$ is some trial function of appropriate type. The optimum $g(t)$ is then the one which minimises

$$E(g) = \int_0^1 \left[f(t) - g(t) \right]^2 dt. \tag{6.1}$$

If, for example, $g(t)$ is a polynomial of degree n,

$$g(t) = \sum_{i=0}^{n} a_i t^i,$$

then $E(g)$ is evaluated as a quadratic function of $a_0, a_1, \ldots, a_n$ and the necessary conditions

$$\frac{dE}{da_j} = 0$$

produce the 'normal equations' which can be solved for the coefficients. Disadvantages of this direct approach are that it does not provide an immediate estimate of the error involved in the approximation, and that the entire computation must be repeated if a more accurate higher degree approximation is required.

If alternatively we assume that $U_0(t), U_1(t), \ldots, U_n(t)$ is an orthonormal basis of the space of polynomials of degree $\leq n$ such that

$$\int_0^1 U_i(t) U_j(t)\, dt = \delta_{ij} = \begin{cases} 1 & i = j \\ 0 & i \neq j \end{cases},$$

then the optimum approximation $g(t)$ is expressible as

$$g(t) = \sum_{i=0}^{n} b_i U_i(t).$$

Substituting into (6.1) and using orthogonality gives

$$E(g) = \int_0^1 \big(f(t)\big)^2 dt + \sum_{i=0}^{n}\left(\int_0^1 f(t) U_i(t)\, dt - b_i\right)^2 - \sum_{i=0}^{n}\left(\int_0^1 f(t) U_i(t)\, dt\right)^2 \quad (6.2)$$

Clearly (6.2) is minimised when

$$b_i = \int_0^1 f(t) U_i(t)\, dt$$

and the minimum value in this case is:

$$E(g_0) = \int_0^1 \big(f(t)\big)^2 dt - \sum_{i=0}^{n} b_i^2. \quad (6.3)$$

Using this orthogonal polynomial approach the coefficients b_i can be directly evaluated as integrals, and the corresponding mean square error is immediately available from equation (6.3). If a higher degree approximation is required it is necessary to evaluate only the higher order coefficients rather than repeat the entire computation. The disadvantage of the simple orthogonal polynomial method based upon Legendre polynomials is that approximation errors are as likely at the end points $t = 0, t = 1$ as at any other point. In this respect it corresponds to the use of standard Chebyshev polynomials for degree reduction.

6.3   Constrained orthogonal polynomials

Following the philosophy of [15] we consider the construction of orthogonal polynomials satisfying particular constraints at $t = 0$ and $t = 1$. In the following analysis only simple constraints $f(0) = f(1) = 0$ are considered but the method can be readily extended to accommodate derivative constraints. Fortunately, the set of polynomials of degree $\leq n$ with $f(0) = f(1) = 0$ forms a proper subspace of the space of real polynomials. A simple basis is

$$t(t-1), t^2(t-1), \ldots, t^{n-1}(t-1).$$

With inner product

$$\left[f(t), g(t)\right] = \int\limits_0^1 f(t)g(t)\, dt,$$

this is not an orthogonal basis, but can be orthonormalised using the Gram-Schmidt algorithm to produce an orthogonal basis $U_1(t), \ldots, U_{n-1}(t)$ of increasing degree.

In the simple case $n = 3$ the orthonormal basis is

$$U_1(t) = \sqrt{30}\, t(t-1)$$

$$U_2(t) = \sqrt{210}\, t(t-1)(2t-1).$$

The following sections will describe how these functions can be used to find best (least squares) cubic approximations to simple parametric curves and surfaces.

For more general constraints and higher degree polynomials a recurrence relation based upon the Gram-Schmidt algorithm provides a more convenient method of generating the required orthogonal basis. If derivative constraints of orders $l - 1$ and $k - 1$ are required at the end points, the generating algorithm is:

$$a_0 = 0$$

$$f_1(t) = t^l(1-t)^k,$$

for $i > 0$

$$a_i^2 = \int_0^1 \left(f_i(t)\right)^2 dt,$$

$$u_i(t) = f_i(t)/a_i,$$

$$b_i = \int_0^1 t\left(u_i(t)\right)^2 dt,$$

$$f_{i+1}(t) = tu_i(t) - b_i u_i(t) - a_i u_{i-1}(t). \tag{6.4}$$

This algorithm can be used to generate the explicit orthonormal basis functions but it has been found more convenient to store the values a_i, b_i and use formula (6.4) in the evaluation of the polynomials when required. This technique has been successfully implemented to produce families of simple and derivative constrained orthogonal polynomials of degree up to 21.

6.4 Curve approximation

The basic problem is that of obtaining for a specified parametric curve $\mathbf{r}(t)$ for $0 \le t \le 1$, an approximating curve $\mathbf{s}(t)$ with $\mathbf{s}(0) = \mathbf{r}(0)$, $\mathbf{s}(1) = \mathbf{r}(1)$ and giving an 'optimum fit' at all other points. In the Chebyshev economisation method described in Chapter 5, the minimax criterion was used as a measure of optimality; for the orthogonal polynomial method the least squares criterion is chosen. The function to be minimised is:

$$\Phi(s) = \int_0^1 \left(\mathbf{r}(t) - \mathbf{s}(t)\right)^2 dt.$$

For this minimisation the three vector components of $\mathbf{r}$ can be treated independently. The resulting approximating curve will be one which is close to the original curve and has a similar parametrisation, it is not necessarily the closest fitting curve of a given degree if parametrisation is ignored.

Since the orthogonal functions have roots at $t = 0$ and $t = 1$ the function $\mathbf{r}(t)$ is modified by subtracting the straight line joining $\mathbf{r}(0)$ to $\mathbf{r}(1)$, giving

$$\mathbf{r}_0(t) = \mathbf{r}(t) - \Big[\mathbf{r}(0) + t\big(\mathbf{r}(1) - \mathbf{r}(0)\big)\Big].$$

The function $\mathbf{r}_0(t)$ then has $\mathbf{r}_0(0) = \mathbf{r}_0(1) = 0$ and can be approximated as

$$\mathbf{s}_0(t) = \sum_{i=1}^{n-1} \mathbf{b}_i U_i(t)$$

where

$$\mathbf{b}_i = \int_0^1 \mathbf{r}_0(t) U_i(t)\, dt. \tag{6.5}$$

The mean square error in this approximation is

$$E = \int_0^1 \big(\mathbf{r}_0(t)\big)^2 dt - \sum_{i=1}^{n-1} \mathbf{b}_i^2. \tag{6.6}$$

The corresponding approximation to $\mathbf{r}(t)$ of degree n is then

$$\mathbf{s}(t) = \mathbf{s}_0(t) + \mathbf{r}(0) + t\big(\mathbf{r}(1) - \mathbf{r}(0)\big)$$

with $\mathbf{s}(0) = \mathbf{r}(0)$, $\mathbf{s}(1) = \mathbf{r}(1)$ and mean square error given by (6.6). If the errors indicated by (6.6) are unacceptably large they can be reduced either by increasing the degree of the approximation or by subdividing the curve into segments; both these techniques have been used in the example below.

6.4.1 Curve approximation example

A circle is one of the simplest examples of a curve which cannot be precisely represented in parametric polynomial form. The equation

$$\mathbf{r}(t) = 5\cos\,(2t)\,\mathbf{i} + 5\sin\,(2t)\,\mathbf{j} \quad \text{for } 0 \le t \le 1,$$

represents a segment of a circle centered at the origin and radius 5. The corresponding function $\mathbf{r}_0(t)$ is thus (to 6 decimal place accuracy)

$$\mathbf{r}_0(t) = \Big(5\cos 2t - 5 + 7.0807342t\Big)\mathbf{i} + \Big(5\sin 2t - 4.546487t\Big)\mathbf{j}.$$

Using equation 6.5 to evaluate the coefficients of $U_1(t)$ and $U_2(t)$ gives

$$\mathbf{b}_1 = (-0.89570844, -1.39498306, 0),$$

$$\mathbf{b}_2 = (0.18098569, -0.11620968, 0),$$

with mean square error, given by equation 6.6, of 5.7 x 10^{-4}.

In this simple case the required integrals can be evaluated analytically but for more general problems some form of numerical integration is required. The accuracy of the approximation can be improved either by increasing the degree or subdividing the curve. Increasing the degree to 4 leaves the first two coefficients unchanged, introduces the coefficient $\mathbf{b}_3 = (0.01290811, 0.02012304, 0)$, and reduces the mean square error to 2.6 x 10^{-6}.

Alternatively subdividing the curve into two segments and still using cubic approximation reduces the mean square error to 2.4 x 10^{-6} but tangent continuity at the midpoint is lost.

Figures 32 and 33 show respectively the original curve with the single span cubic approximation, and the original curve with the quartic approximation. In each case the approximation curve is denoted by a dashed line. As in the quartic case the discrepancies between the subdivided cubic approximation and the original curve can hardly be detected by eye.

6.5   Surface approximation

For an approximation to a patched parametric surface to be effective it is not sufficient merely to compute surfaces which approximate closely to each patch, since the reconstructed surface might exhibit unwanted discontinuities at patch to patch boundaries. It is of course unrealistic to expect the low degree approximation surface to match exactly the boundaries of the original surface but any approximation errors must be consistent with those made on the boundary of the adjacent patch. Similar continuity considerations apply if it is necessary to subdivide a surface in order to obtain a sufficiently close approximation of the given degree.

The above constraints make it impossible to simply obtain the required parametric surface approximation as a tensor product of simple unconstrained orthogonal polynomials. The least acceptable constraint requires C^0 continuity from patch to patch in a piecewise parametric surface approximation. This can be achieved as a three stage process using the constrained orthogonal polynomials described in Section 6.3. Given a parametric surface $\mathbf{r}(u,v)$, $0 \leq u \leq 1$, $0 \leq v \leq 1$ the objective is

Figure 32. Circular arc and cubic approximation (dashed line)

to construct a parametric polynomial approximating surface $s(u, v)$ in such a way that any adjacent surface will have a consistent approximation along the common boundary ($u = 0, v = 0, u = 1$ or $v = 1$). The approximation should be the one that minimises the objective function

$$\Phi(s) = \int\limits_0^1 \int\limits_0^1 \Big[r(u, v) - s(u, v) \Big]^2 \, du \, dv.$$

In the first stage of the approximation the methods described in Section 6.4 are used to obtain approximation curves $a_0(u), a_1(u), b_0(v), b_1(v)$ to the boundaries $r(u, 0), r(u, 1), r(0, v), r(v, 1)$ respectively. The constraints included will ensure that these boundary approximation curves will intersect at the patch corners. In the second stage the boundary approximation curves are used to construct a Coons' surface:

$$c(u, v) = (1 - v)a_0(u) + va_1(u) + (1 - u)b_0(v) + ub_1(v)$$

$$-(1 - u)(1 - v)r(0, 0) - u(1 - v)r(1, 0) - v(1 - u)r(0, 1) - uvr(1, 1). \qquad (6.7)$$

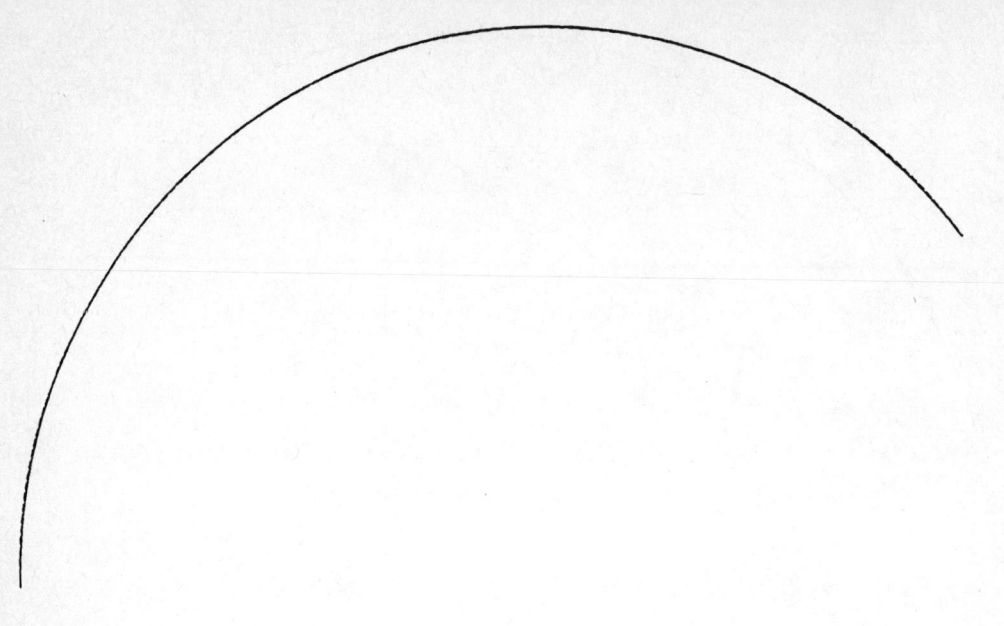

Figure 33. Circular arc and quartic approximation (dashed line)

Then $\mathbf{c}(u,v)$ is a polynomial parametric surface of the required degree which will coincide with $\mathbf{r}(u,v)$ at the four corner points and fit as closely as possible along the boundary curves. There may however be significant errors at interior points. These errors are minimised if, as the final stage of the approximation process, the difference function

$$\mathbf{d}(u,v) = \mathbf{r}(u,v) - \mathbf{c}(u,v) \tag{6.8}$$

is constructed and approximated as a tensor product of constrained orthogonal polynomials

$$\mathbf{d}(u,v) \approx \sum_{i=1}^{n-1}\sum_{j=1}^{n-1} \mathbf{a}_{ij}U_i(u)U_j(v) = \mathbf{d}_A(u,v). \tag{6.9}$$

The coefficients in (6.9) are obtained by integration as:

$$\mathbf{a}_{ij} = \int_0^1 \int_0^1 \mathbf{d}(u,v)U_i(u)U_j(v)\,du\,dv. \tag{6.10}$$

The final surface approximation, of degree n in u and in v, is obtained by adding the correction $\mathbf{d}_A(u,v)$ to $\mathbf{c}(u,v)$ giving:

$$\mathbf{s}(u,v) = \mathbf{c}(u,v) + \mathbf{d}_A(u,v). \tag{6.11}$$

As with the curve approximation the method provides precise values for the mean error in the approximation

$$E = \int_0^1 \int_0^1 \left(\mathbf{d}(u,v)\right)^2 du\, dv - \sum_{i=1}^{n-1}\sum_{j=1}^{n-1}\left(\mathbf{a}_{ij}\right)^2. \tag{6.12}$$

Strictly (6.12) gives the error in using $\mathbf{d}_A(u,v)$ as an approximation to $\mathbf{d}(u,v)$ but this is identical to the discrepancy between $\mathbf{r}(u,v)$ and $\mathbf{s}(u,v)$.

6.5.1 Simple surface approximation example

A simple example of a non-polynomial parametric surface is a torus. A patch of this surface has parametric equation

$$\mathbf{r}(u,v) = \sin v \left(10 + 5\cos u\right)\mathbf{i} + \cos v \left(10 + 5\cos u\right)\mathbf{j} + 5\sin u\mathbf{k}$$

for $(u,v) \in [0,1] * [0,1]$. This is a torus centered on the z axis with major radius 10 and minor radius 5. For this surface the boundary curves, which are all circular curves are:

$$\mathbf{r}(u,0) = (10 + 5\cos u)\mathbf{j} + 5\sin u\,\mathbf{k}$$
$$\mathbf{r}(u,1) = (10 + 5\cos u)(\sin 1\,\mathbf{i} + \cos 1\mathbf{j}) + 5\sin u\,\mathbf{k}$$
$$\mathbf{r}(0,v) = 15\sin v\,\mathbf{i} + 15\cos v\,\mathbf{j}$$
$$\mathbf{r}(1,v) = (10 + 5\cos 1)(\sin v\,\mathbf{i} + \cos v\,\mathbf{j}) + 5\sin 1\,\mathbf{k}$$

The technique described in section 6.4 is used to find the best cubic approximations $\mathbf{a}_0(u), \mathbf{a}_1(u), \mathbf{b}_0(v), \mathbf{b}_1(v)$ respectively which interpolate the end points. Using equation 6.6 the maximum mean square error in these approximations is 2.2×10^{-5} for boundary curve $\mathbf{b}_0(v)$. The Coons patch $\mathbf{c}(u,v)$ defined by equation 6.7 is then constructed and subtracted from $\mathbf{r}(u,v)$ to define the difference surface $\mathbf{d}(u,v)$.

Equation 6.10 is then used to find the vector coefficients for the orthogonal approximation $\mathbf{d}_A(u,v)$ to this difference surface. In this simple case the integrals can be evaluated analytically to give:

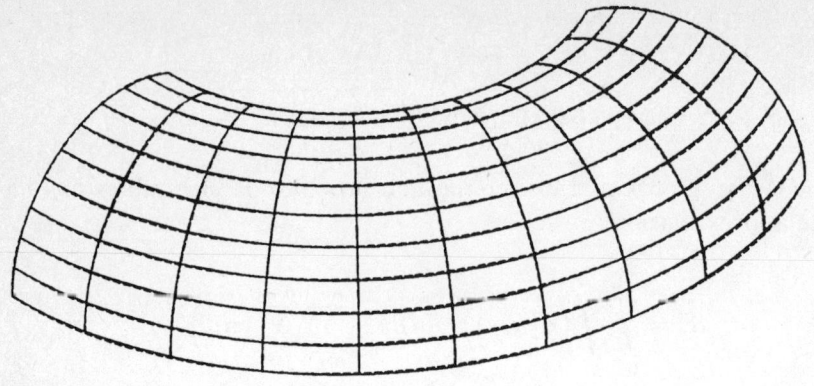

Figure 34. Bi-cubic orthogonal approximation (dashed) to torus section

$$\mathbf{a}_{11} = (-0.016713, 0.030593, 0), \quad \mathbf{a}_{12} = (-0.001941, -0.0001065, 0)$$

$$\mathbf{a}_{21} = (0.0005793, -0.0001065, 0) \quad \mathbf{a}_{22} = (0.0000673, 0.0000368, 0)$$

Using equation 6.12 the mean square error in the final approximation $\mathbf{c}(u, v) +$ $\mathbf{d}_A(u, v)$ is approximately 2.3 x 10^{-5}. From this and the coefficients $\mathbf{a}_{ij}$ the mean square error in $\mathbf{c}(u, v)$ can be estimated to be 1.24 x 10^{-3} showing that the final approximation is much closer to the surface than the intermediate Coons' patch. Increasing the degree of the orthogonal approximation to 4 reduces the mean square errors to a maximum of 2.1 x 10^{-8} on the boundaries and 2.2 x 10^{-8} for the entire surface approximation.

Figure 34 displays the bi-cubic approximation to a 2 radian by 2 radian section of the torus of major radius 10 and minor radius 5. The approximation (shown dashed) is superimposed on the original torus section. Figure 35 shows the corresponding bi-quartic approximation. Here the discrepancy cannot be detected by eye.

6.6 More general curve approximation problems

In section 6.4 it was shown that the essential computational steps in evaluating an approximating curve and estimating the mean square error in the approxima-

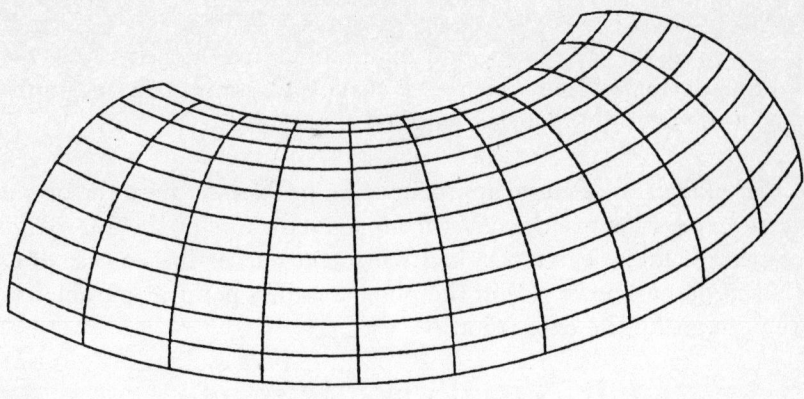

Figure 35. Bi-quartic orthogonal approximation (dashed) to torus section

tion involved the evaluation of the integrals given by (6.5) and (6.6). If $r(t)$, and consequently $r_0(t)$ is a simple parametric function these integrals can be evaluated analytically. When $r(t)$ is more complex the integrals must be evaluated by a suitable numerical integration technique such as Gaussian quadrature. The oscillatory nature of the orthogonal basis functions means that these integrals have to be evaluated carefully with the integration range sub-divided, particularly if an approximation of a comparatively high degree is required. Once this has been implemented the implication is that an orthogonal approximation and error estimate can be found for any parametric curve $r(t)$ which can be evaluated at the integration points. It is not a requirement that a simple explicit formulation for $r(t)$ should be available.

Using the algorithm defined by (6.4) families of orthogonal polynomials can be constructed to meet any required derivative constraints at the left and right hand ends of curve segments. In order to implement these constraints $r_0(t)$ must be redefined as the difference between $r(t)$ and any other parametric function $q(t)$ of acceptable degree which matches $r(t)$ and the appropriate number of derivatives at the ends $t = 0, t = 1$. In practice it is convenient to construct $q(t)$ as a Hermite or Bézier polynomial curve. If, as is frequently the case, the real problem is to construct a polynomial approximation of a specified degree which is within a given tolerance of the original curve the error estimate can be used to decide when it is necessary to subdivide the curve and approximate as 2 or more pieces. In this case the approximation will be a piecewise curve and a minimum of first derivative continuity will be generally required between the segments.

Two particular examples of curve approximation problems where the original curve has no explicit expression are the offset from a given curve or the approximation by a single segment of high degree of a multi-segment curve. In the case of a tapered offset to a given planar curve $\mathbf{r}(t)$ in the plane $z = 0$, a point at parameter t on the offset curve is given by the expression

$$\mathbf{R}(t) = \mathbf{r}(t) + (\mathbf{k} * \mathbf{T})(\mathbf{d}_0(1 - t) + t\mathbf{d}_1)$$

where $\mathbf{T}$ is the unit tangent vector in the direction and $\mathbf{d}_0$ and $\mathbf{d}_1$ are the initial and final offset distances. It should be noted that even in the simple case, where $\mathbf{r}(t)$ is of polynomial form and $\mathbf{d}_0 = \mathbf{d}_1$, the offset curve $\mathbf{R}(t)$ is rarely a simple polynomial curve. The orthogonal approximation method can be applied to find an approximation to $\mathbf{R}(t)$ of any specified degree and, with subdivision if necessary, within any required tolerance.

Figure 36 shows the result of applying this process to compute the approximation of degree 5 to the standard offset ($\mathbf{d}_0 = \mathbf{d}_1$) to a two segment cubic Bézier curve. The approximating curve has been converted to Bézier form and both the original and offset control points are displayed. Figure 37 shows the result of applying this process to compute the approximation of degree 4 to the tapered offset to a two segment cubic Bézier curve. The offset curve consists of four segments.

In the case of a composite curve the curve first has to be re-parametrised to give a parameter range of $0 \leq t \leq 1$ for the entire curve and the detailed evaluation of $\mathbf{r}(t)$ is then dependent upon first determining the correct curve segment. This process is appropriate when it is required to transfer spline curves to a system which permits high degree curves but not multi-segment curves. Figures 38 and 39 show a three segment Bézier curve of degree 4 and the corresponding approximation curves of degrees 7 and 11 respectively. The control polygon of the original curve is also displayed in Figure 38.

6.7　More general surface approximation problems

As for curves, the orthogonal approximation of surfaces can be reduced to the evaluation of a number of integrals. Using numerical techniques this reduces to a requirement to be able to evaluate $\mathbf{r}(u, v)$ at discrete points on the surface and hence this technique can be applied to approximate any parametric surface which can be

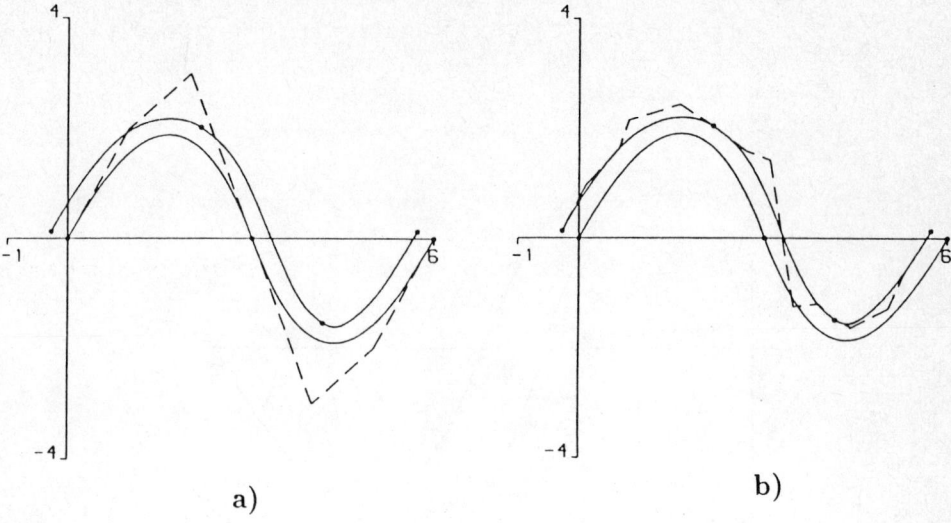

Figure 36. Offset approximation of degree 5 to cubic Bézier curve:
a) with original control polygon   b) with offset control polygon

evaluated. If more general derivative constraints are required to be met at the patch corners and along the surface boundary the method described in section 6.5 requires to be generalised. The simply constrained orthogonal basis functions are replaced with the appropriate derivative constrained basis functions. The 4 boundary curves are then approximated in the usual way. The simple Coons patch $c(u,v)$ defined in (6.7) must now be replaced by any surface $c_1(u,v)$ which interpolates the 4 boundary curves and conforms to the specified constraints at the corners and along the edge curves of the surface. In the case of first derivative constraints $c_1(u,v)$ can either be a more generalised Coons' patch using cubic blending functions in place of the linear ones in (6.7), or could be a Bézier surface in which the control points at the edge are fully determined by the boundary approximation curves and some of the inner control points are determined from the corner and cross boundary derivatives. In either case $d(u,v) = r(u,v) - c_1(u,v)$ will be a surface with zero value and zero derivatives at the corners which can be approximated by the derivative constrained orthogonal polynomials to give $d_A(u,v)$. The final approximation surface $d_A(u,v) + c_1(u,v)$ will coincide with $r(u,v)$ at the corners, and along the boundaries will be coincident with the boundary curve approximations with the appropriate cross boundary derivatives.

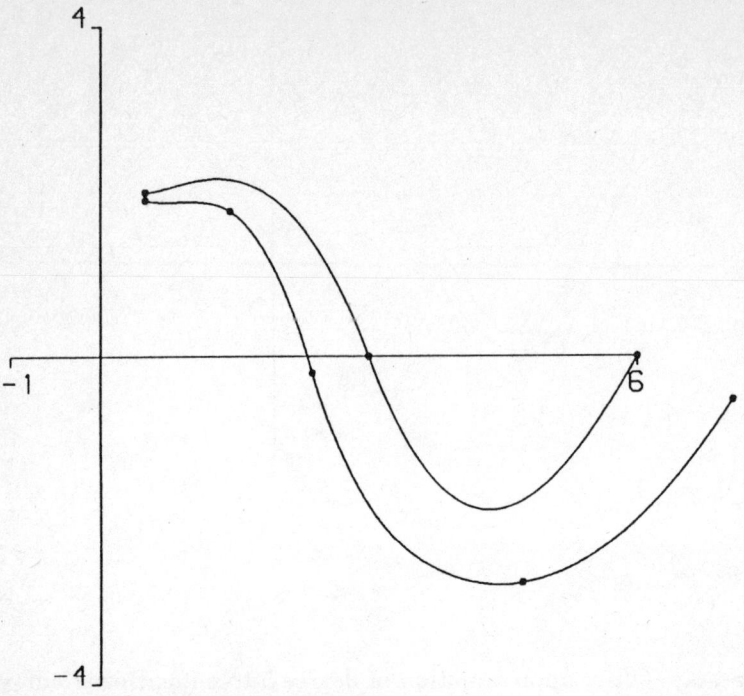

Figure 37. Tapered offset approximation of degree 4 to cubic Bézier curve

The orthogonal approximation technique can be applied to a wide range of surface approximation problems. These include degree reduction problems, the approximation of rational surfaces by polynomial surfaces, the approximation of offset and other forms of procedurally defined surfaces and the approximation of a composite patched surface by a single patch surface of higher degree. As with curve approximations the approximation error can be monitored by computing the mean square error and, if too large, a better approximation can be obtained by increasing the degrees or by sub-dividing the surface into patches.

The sub-division strategy is more complex for surfaces since available options at any stage are subdivision in the u range or subdivision in the v range or both. Any such subdivision will propagate across the entire surface and, with an inappropriate strategy, a very large number of patches can be produced. Using the mean square errors from the boundary curve approximations it is a simple matter to determine whether subdivision in u, or v is necessary to achieve the desired accuracy along the boundaries. If the boundary approximations are satisfactory but there is an unacceptably large error in the interior of the surface the choice of whether to sub-divide in u, or subdivide in v is less obvious. Two possible methods to determine this are

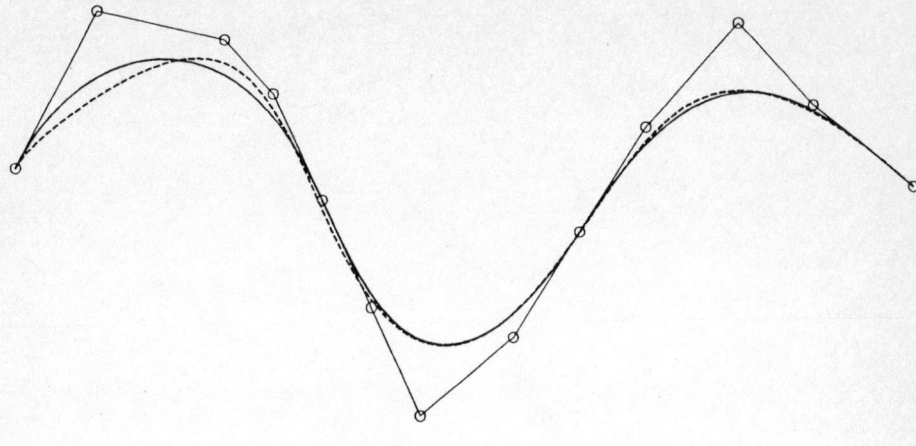

Figure 38. Piecewise low degree (4) to high degree (7) approximation

either to re-examine the boundary errors and, even though these are satisfactory, to subdivide in the parameter corresponding to the largest boundary error, or to compute the approximation of one degree higher and compare the coefficients produced. With the second strategy the subdivision should be made in the parameter producing the larger coefficients. A simple, but effective strategy which has been used to date is to make the subdivision, when required, at the mid-point of the current parameter range.

As an example of approximation to an offset surface Figure 40 shows the result of using the orthogonal approximation method to find the simply constrained approximation of degree 4 in u and v to the offset to part of a toroidal surface. The mean square error in this approximation was approximately 2.5×10^{-8} for a surface with an overall diameter of approximately 25.

A less orthodox application of orthogonal approximation helps to provide a solution to the problem of communicating a trimmed surface, that is a surface with a superimposed boundary curve, to a CAD system without a trimming facility. This transfer is only possible if the trimmed part of the original surface can be reconstructed so that it appears to the receiving system to be a simple parametric surface. If the original surface is a patched piecewise polynomial surface the prob-

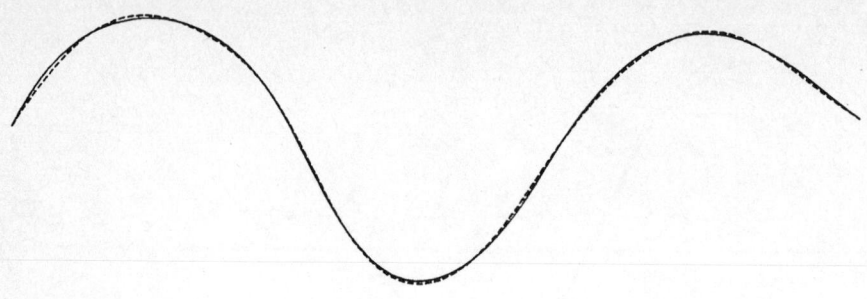

Figure 39. Piecewise low degree (4) to high degree (11) approximation

lem is simplified locally to that of being able to replace one or more of the original patch boundaries by boundaries which correspond to curves in u-v parameter space. The replacement patch must then be re-constructed so that the natural boundaries include the curved boundary. Using the orthogonal approximation method this re-construction is achieved as a two stage process. In the first stage the new boundary curve of the original patch is identified as a curve in (parametric) u, v space and a continuous mapping

$$u = u(U, V), \quad v = v(U, V)$$

is found which transforms the unit square $0 \leq U \leq 1$, $0 \leq V \leq 1$ in (U, V) space to the trimmed region in (u, v) space on the original patch. Figure 41 shows the result of such a mapping superimposed on the original (u, v) space patch, this mapping is displayed by constant parameter lines at intervals of 0.25. If $\mathbf{r}(u, v)$ is the original surface the new surface patch is defined by the equation

$$\mathbf{R}(U, V) = \mathbf{r}(u(U, V), v(U, V)), \quad 0 \leq U \leq 1, 0 \leq V \leq 1.$$

The orthogonal approximation method is then applied in the second stage of this process to find an approximation of the required degree to the surface $\mathbf{R}(U, V)$.

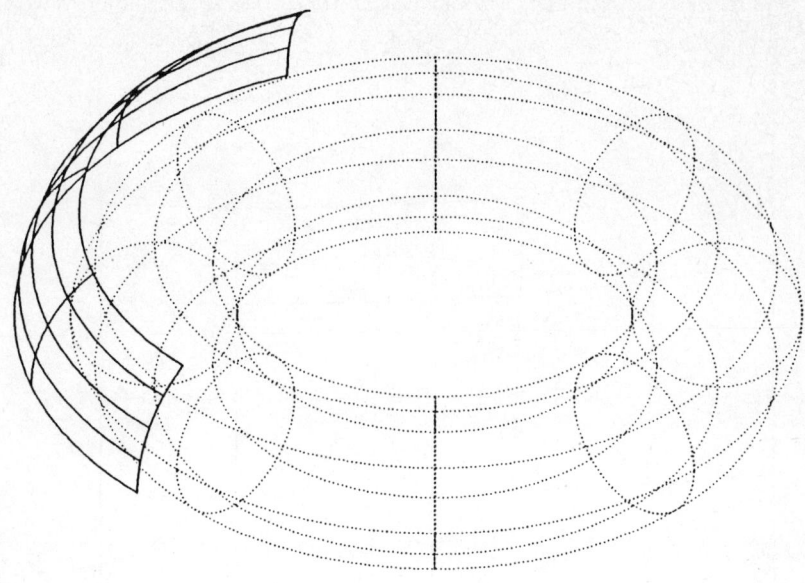

Figure 40. Orthogonal approximation of degree 4 by 4 to part of offset from toriodal surface

This approximation will include approximations to the revised boundary curves. In the example illustrated in Figures 41 and 42 the trimming curve corresponds to the boundary $V = 1$. Figure 42 shows the new trimmed surface patch $\mathbf{R}(U, V)$ surrounded by other patches which have still to be trimmed and uses the original parametrisation to generate the wire frame display. Using this method the maximum deviation of the reconstructed trimmed surface patch from the original surface and boundary curve was less than 1×10^{-3} mm.

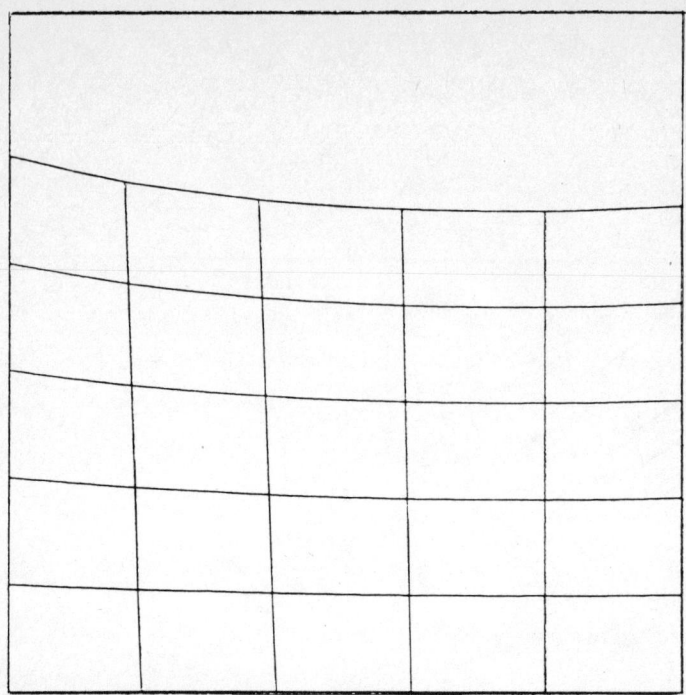

Figure 41. Mapping $u = u(U, V)$, $v = v(U, V)$ superimposed on (u, v) square

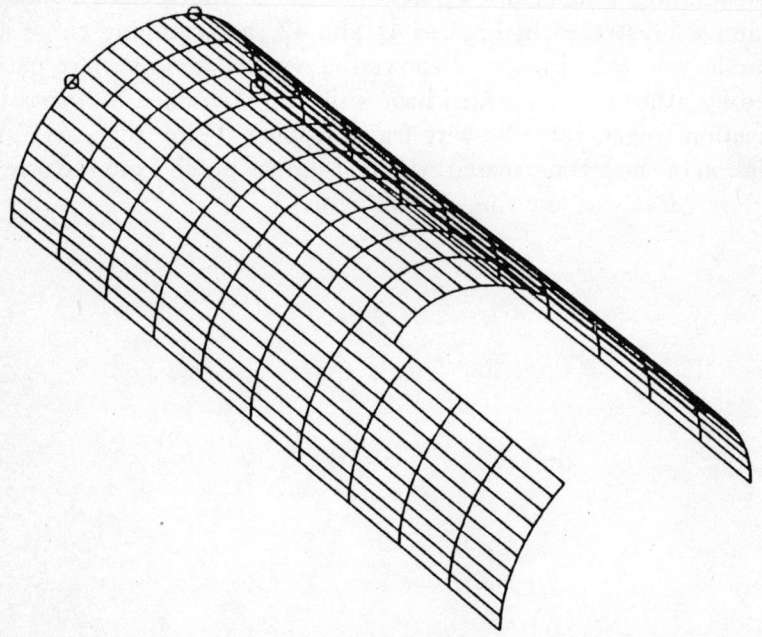

Figure 42. Bi-cubic patch approximation to trimmed surface using (u, v) mapping combined with orthogonal approximation

References

[1] Boehm, W: Inserting new knots into B-spline Curves,
Computer Aided Design, 12, 1980, pp 199-201.

[2] Boehm, W: Generating the Bézier points of B-spline Curves and Surfaces,
Computer Aided Design, 13, 1981, pp 365-366.

[3] Boehm, W et al: A Survey of Curve and Surface Methods in CAGD,
Computer Aided Geometric Design, 1, 1984, pp 1-60.

[4] Boehm, W: On the Efficiency of Knot Insertion Algorithms,
Computer Aided Geometric Design, 2, 1985, pp 141-143.

[5] de Boor, C: On Calculating with B-splines,
Journal of Approximation Theory 6, 1972, pp 50-62.

[6] de Boor, C: A Practical Guide to Splines,
Springer-Verlag , 1978.

[7] Chang, G and Wu, J: Mathematical Foundations of Bézier's Technique,
Computer Aided Design, 13, 1981, pp 133-136.

[8] Cohen, E, Lyche, T and Riesenfeld, R: Discrete B-splines and Subdivision
Techniques in CAGD and Computer Graphics,
Computer Graphics and Image Processing, 14, 1980, pp 87-111.

[9] Cohen, E, Riesenfeld, R: General Matrix Representations for Bézier and
B-spline Curves, *Computers in Industry*, 3, 1982, pp 9-15.

[10] Cox, M. G: The Numerical Evaluation of B-splines
National Physical Laboratory Report DNAC 4, 1971.

[11] Farin, G: Curves and Surfaces for Computer Aided Design,
Academic Press, 1988.

[12] Farouki, R.J and Rajan, V.J: Algorithms for polynomials in Bernstein
form, *Computer Aided Geometric Design*, 5, 1988

[13] Farouki, R.J and Rajan, V.J: On the numerical condition of polynomials
in Bernstein form, *Computer Aided Geometric Design*, 4, 1987

[14] Initial Graphics Exchange Specification (IGES), Version 3.0
National Bureau of Standards, Gaithersburg, Maryland, USA, April, 1986.

[15] Lachance, M.A, Saff, E.B and Varga, R.S: Bounds on incomplete polynomials
vanishing at both end points of an interval, in Constructive Approaches
to Mathematical Models, Academic Press, 1979.

[16] Lachance, M.A: Chebyshev economization for parametric surfaces,
Computer Aided Geometric Design, 3, 1988, pp 195-208.

[17] Lyche, T, Cohen, E and Morken K: Knot Line Refinement Algorithms for
Tensor Product B-spline Surfaces,
Computer Aided Geometric Design, 2, 1985, pp 133-139.

[18] Lyche, T and Morken, K: Making the Oslo Algorithm more efficient,
Siam Journal of Numerical Analysis, 23, 3, 1986.

[19] Prautzsch, H: Degree elevation of B-spline curves,
Computer Aided Geometric Design, 2, 1984, pp 193-198.

[20] Specification of CAD*I Neutral File for CAD Geometry, Version 3.3
E.G Schlechtendahl (editor), ESPIRIT Research Reports, Vol 1, CAD*I,
Springer-Verlag, Heidelberg, 1988.

[21] Standard for the Exchange of Product Definition Data (STEP)
International Standards Organisation, DP10303, 1989

[22] Standard d'Exchange et de Transfert Specifications (SET), Revision 1.1,
Aerospatiale, France, March, 1984.

[23] Verband der Automobilindustrie - Flächen-Schnittstelle (VDA-FS),
Version 2.0, Karlsruhe, Germany, 1986.

APPENDIX A: THE NEUTRAL FILE CHECK SYSTEM

Principal authors: H. Scheder, D. Trippner

A.1   The necessity of data exchange software

As mentioned in chapter 1 there is a great need for data exchange software in industrial applications. This is not only to have tools for the verification of data which is to be exchanged between different systems at BMW or between BMW and its suppliers, but also to facilitate a comprehensive test of pre- and post-processors provided by CAD/CAM system vendors. For this reason a complete test-methodology was developed and introduced as described in chapter 1. The CAD data exchange software tools which are part of this methodology will be described in this Appendix.

A.2   The IGES tools

The following IGES tools have been developed within the CAD*I project for a quick and reliable check of IGES files produced by the systems pre-processors in order to determine the origin of data transfer problems. The IGES tools are also used for a data check to make sure that the data which is intended to be sent is correct in terms of file structure in conformity with the neutral format specification.

IGES tools:

-	ICHECK	IGES checker
-	SYNTAX	for a special syntactical analysis
-	POINTER	for a special analysis of pointers
-	STATISTIC	for a statistical analysis
-	ISCOMP	for a comprehensive statistical analysis

The language used for the programming of the IGES tools is FORTRAN 77, complying with the ANSI X3.9-1978 standard, supplemented with the INCLUDE statement.

Implementations of these tools are running on the following operating systems:

* IBM VM/CMS,
* VAX VMS,

* UNIX and

* MS-DOS.

A.2.1 The SYNTAX analysis program

The SYNTAX analysis program checks the correctness of IGES files with respect to:

- the number of parameters of each entity,

- the syntax of file sections, each entry and each parameter,

- the data types (INTEGER/REAL/CHARACTER) of parameters,

- the sequence numbers,

- the value ranges and

- the record numbers of each section.

The messages noted within the error report produced by the syntax program are classified into fatal errors and warnings. A difference between the occurrence of an entity in the data file and the specification of that entity is that it may produce either a fatal error or a warning depending on the CAD system for which the data is intended. For this reason the error table can be edited and adapted to the requirements of any CAD system. A detailed output contains the position of an error indicated by the sequence number. The number of the corrupted parameter can be produced if wanted.

The program supports the IGES entities contained in one of the VDAIS subsets and some common IGES entities not included within the VDAIS specification.

As shown in Figure 43 the SYNTAX analysis program consists of different modules for parsing, scanning and checking the data under the supervision of a control module. The check routines have access to the IGES file by a scanner, for the free formatted entries in the global and the parameter data section, and by a simple formatted read statement for the other sections of the IGES file. For missing or unsuitable global section parameters, default values are used to check the data against. For a semantic check of the contents a special parser routine for each section exists.

The errors recognised and counted are stored in an output file which is used at the end of the program to create a readable error report with detailed messages classified into fatal errors and warnings.

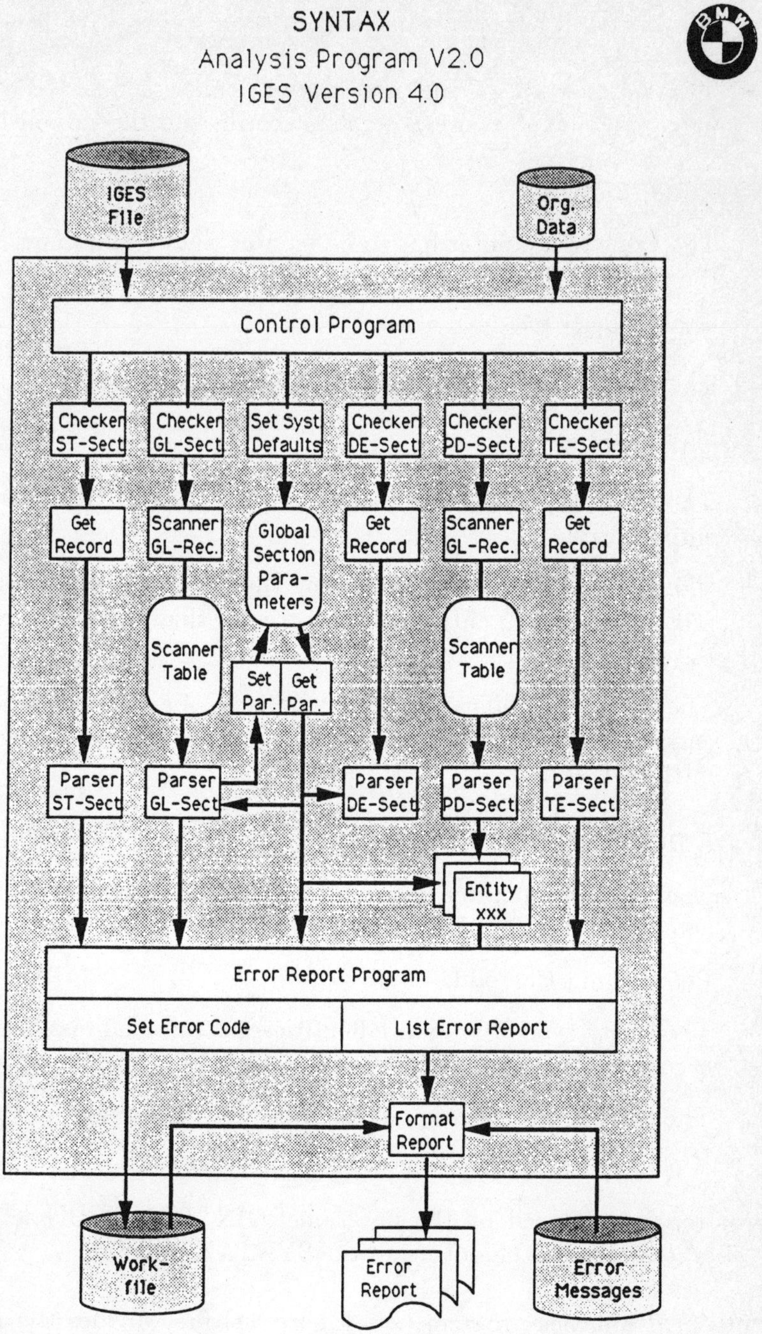

SYNTAX
Analysis Program V2.0
IGES Version 4.0

Figure 43. System design SYNTAX

A.2.2   The POINTER analysis program

The POINTER program checks the correctness of pointers in an IGES file. In detail the following cases of pointers are observed according to the rules listed below.

DE ⟶ DE POINTERS

* The value of a pointer has to be greater than 0 and lower or equal to the sum of DE records minus 1.

* Pointer must be odd.

* The type of the referenced entity has to be admissible.

DE ⟶ PD POINTERS

* The entity type value has to be the same in both DE records for one entity entry.

* The value of the pointer indicating the parameter entry for a given (DE parameter 2) entity must be greater than 0 and lower or equal to the number of PD records.

* The parameter indicating the entity type of a referenced parameter entry in PD section must be identical to the type of the referring entity in the DE section.

PD ⟶ DE POINTERS

* The value of a pointer has to be within the range of 0 and lower or equal to the sum of DE records minus 1.

* Pointers must be odd.

* The entity type of referenced entities has to be admissible and in accordance with the IGES specification.

* The correctness of backpointers from the PD-section up to the DE-section is checked.

The error report produced by the program POINTER and the set of supported entity types is similar to the output of the SYNTAX program.

The POINTER analysis program (see Figure 44) first divides the IGES file into three separate files which allows, within a loop over all entries, a direct access to the entities. Afterwards each pointer relating DE-DE, DE-PD and PD-DE entries

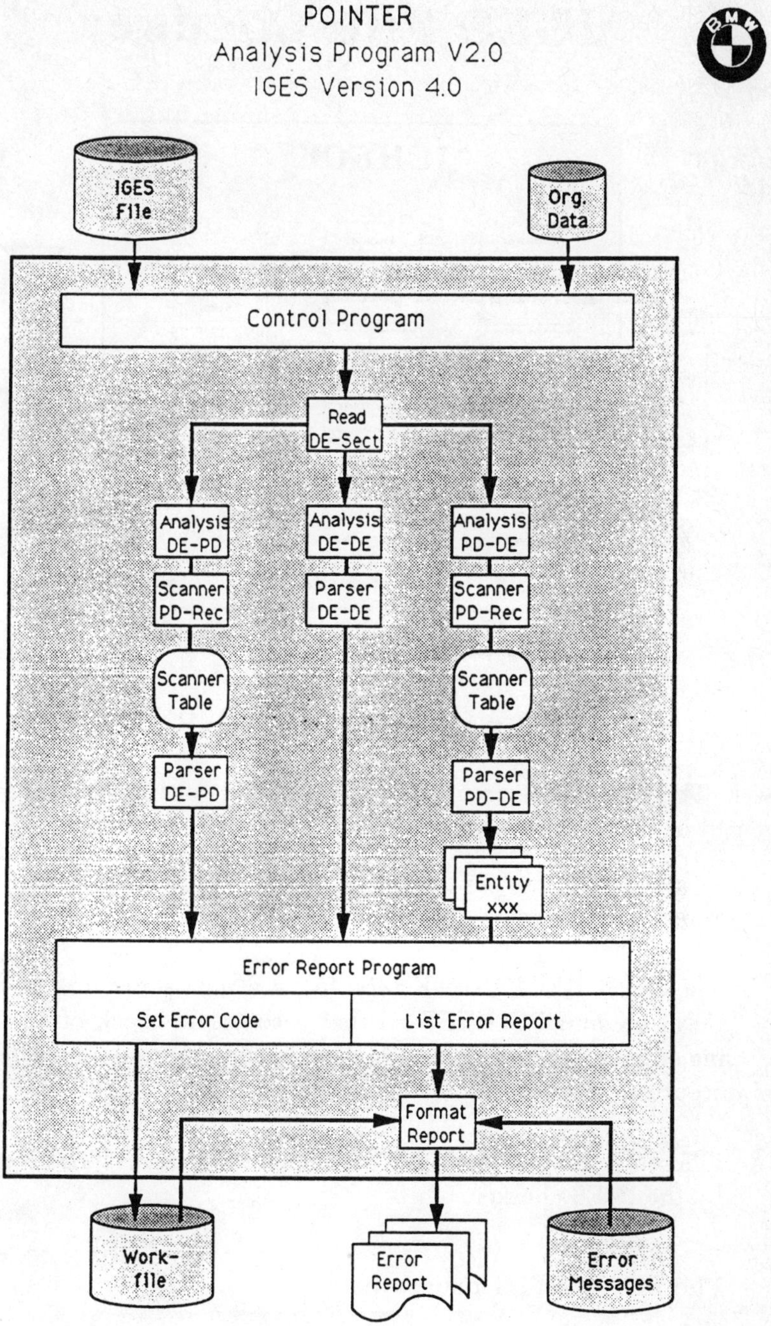

POINTER
Analysis Program V2.0
IGES Version 4.0

IGES
File

Org.
Data

Control Program

Read
DE-Sect

Analysis
DE-PD

Analysis
DE-DE

Analysis
PD-DE

Scanner
PD-Rec

Parser
DE-DE

Scanner
PD-Rec

Scanner
Table

Scanner
Table

Parser
DE-PD

Parser
PD-DE

Entity
xxx

Error Report Program

Set Error Code

List Error Report

Format
Report

Work-
file

Error
Report

Error
Messages

Figure 44. System design POINTER

IGES-File-Checker

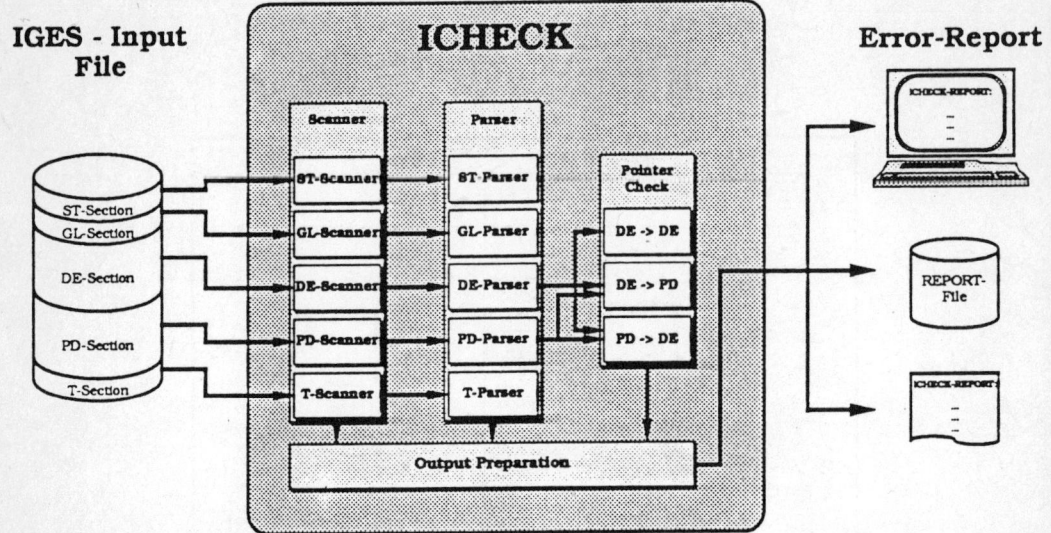

Figure 45. System design ICHECK

is checked. The POINTER program uses a scanner routine similar to the routine implemented in the SYNTAX program to scan the PD-section.

A.2.3 The IGES check program

The IGES checker ICHECK (see Figure 45) is a new program. It combines the programs SYNTAX and POINTER so that a complete check of an IGES file can be done using only one program. A reduction in running time and a redesign of the program output are the benefits of this development.

Today 37 different entity types with 35 form numbers of IGES version 4.0 are supported by the IGES checker.

A.2.4 The STATISTIC program

The STATISTIC program (see Figure 46) gives a short survey of the contents of an IGES file. It is a very useful tool for the support of CAD/CAM data exchange.

* It lists the start section.

Statistic Analysis of IGES-Files

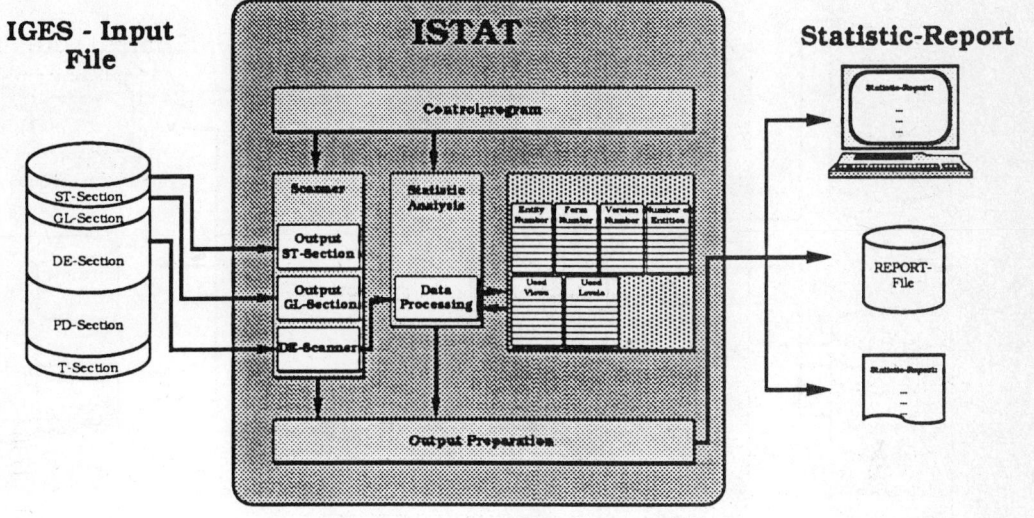

Figure 46. System design STATISTIC

* It lists the contents of the global section in readable form.

* It sets up a statistic of used entity types and forms.

* It sets up a statistic of used attributes like:

    - blank status,

    - subordinate flag,

    - level,

    color,

    - line font and

    - line weight.

* Finally it lists the transformation matrices and view entities referenced.

A.2.5   The IGES statistic comparator program ISCOMP

The ISCOMP program (see Figure 47) is a helpful tool for comparing the contents of two IGES files. It is mainly used for processor testing as described in chapter 1

IGES Statistic Comparator

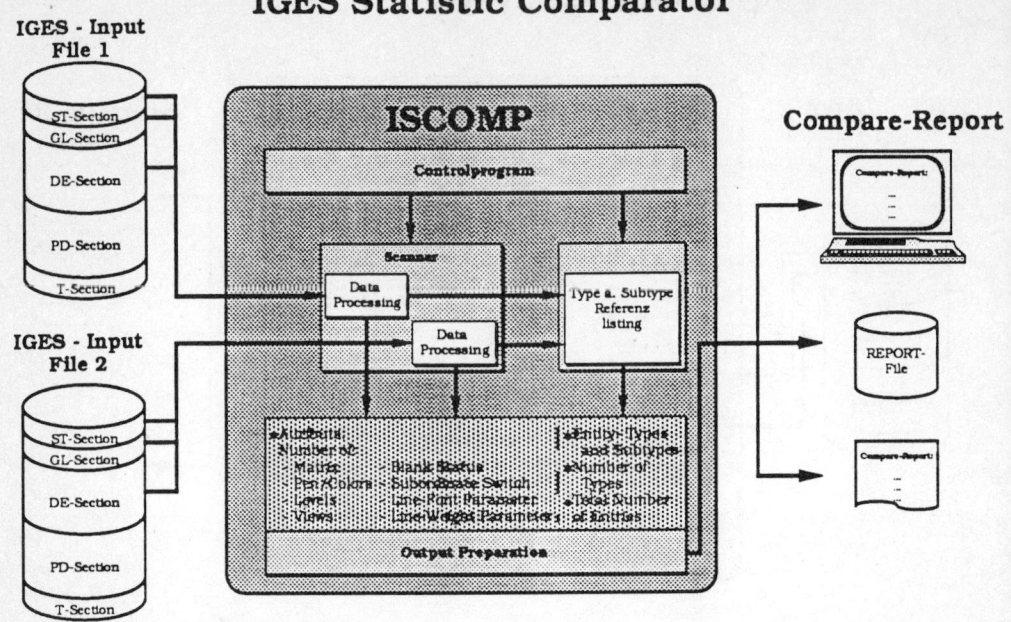

Figure 47. System design ISCOMP

and it produces an output similar to the output of the STATISTIC program, but for two IGES files at the same time in a comparative form.

A.3   The VDAFS ANALYZER

Analogous to the IGES tools a check program for the VDAFS format was developed.

The VDAFS ANALYZER supports the complete entity set specified in version 2.0 of VDAFS and checks the following items:

* Syntax and File conventions

    - syntax and uniqueness of entity type names

    - correctness of separators

    - types of parameters (REAL, INTEGER, NAME)

    - correctness of sequence numbers

    - designation of comment and continuation lines

    - completeness of the file

*   Contents of data

-	structure entities:	begin/end of file
		begin/end of set
		references
-	geometric entities:	parameter type and number
		continuity (CURVE, SURF)
		degree (CURVE, SURF)
		monotony of segment parameters (CURVE, SURF)
		references

In addition, the program sets up a statistical overview of the file contents.

The VDAFS Analyser is written in FORTRAN77, complying with the ANSI X3.9-1978 standard and is running on the following operating systems:

*   IBM VM/CMS,

*   VAX VMS and

*   UNIX.

The VDAFS Analyser consists of several modules as shown in Figure 48. A control program calls to a scanner routine to get the name and type of an entity and control is then passed to a type dependent parser subroutine. Scanner and parser routines count the occurrences of deviations of the VDAFS standard. Finally, a protocol is produced which contains statistical information about the contents of the file as well as a description of the errors which were found.

Analysis of VDAFS-Files

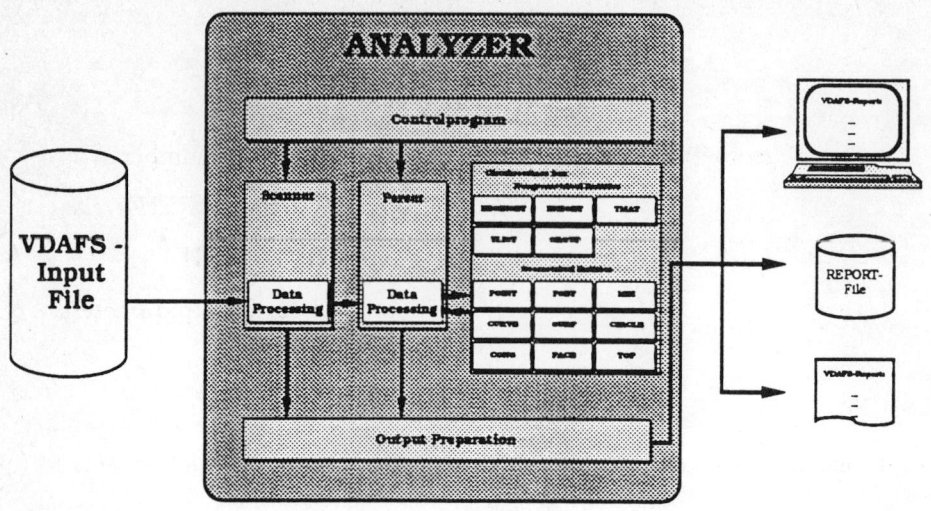

Figure 48. System design VDAFS ANALYZER

APPENDIX B: THE NEUTRAL FILE ADAPTING SYSTEM NFAS

Principal authors: H. Scheder, D. Trippner

B.0   Introduction

NFAS is a system to adapt neutral format data according to requirements determined by the capability of the receiving system and with respect to the applications for which the data is to be used. Hence NFAS must be implemented within the data exchange process, as can be seen in Figure 49, and must be able to run in the background. Although NFAS can be used in an interactive mode it is intended to be used as a batch system. For this reason the commands to control the process have to be stored in a command file which are interpreted at the time of execution.

B.1   System design

The principle which leads to the chosen system design was the strict division of the call interface and the control program which does the mapping between the different representations. The separation of the two parts was forced by the experience that reading and writing of neutral file formats is independent to the functions which have to be executed upon the data. Due to this concept the mapping part of NFAS is nearly independent of a specific neutral file format and in principle is able to work on the base of any CAD data format. Besides the implemented IGES call interface, interfaces for processing VDAFS, STEP or other CAD data formats can be developed.

Moreover, the portability of the system and the necessity to examine about 40,000 statements of source code was taken into consideration. The system was therefore designed in a highly modular form with a clear interface between a format independent NFAS control program and a format dependent IGES call interface based on IGES version 4.0.

B.1.1   NFAS control program

The NFAS control program is partitioned into groups of modules. Each group processes a homogeneous set of tasks. The communication between module groups is based exclusively on subroutine and function calls and, if unavoidable, on common blocks.

Examples for the Use of NFAS

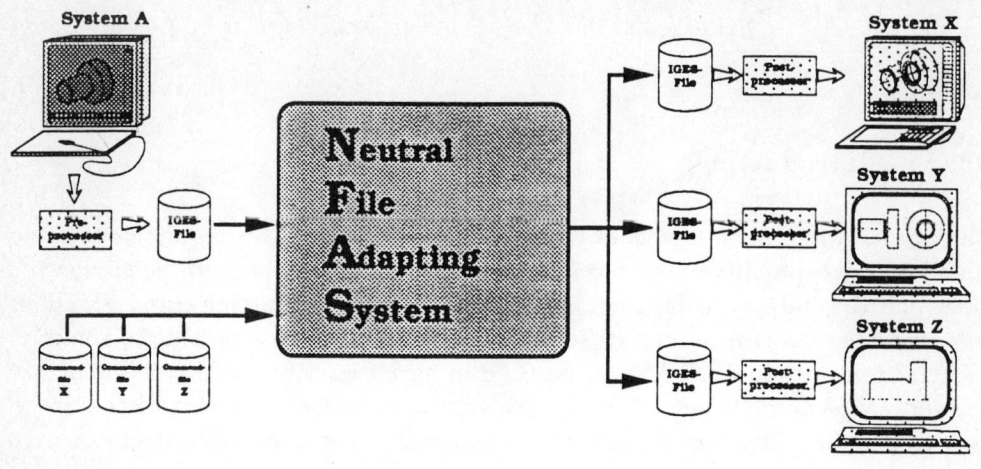

Figure 49. The use of NFAS within the data exchange process

The groups of modules are listed in the following tables with those NFAS-specific first and those independent from NFAS second.

(a) NFAS mainline

These modules are the mainline of the system. It contains the main program loop which reads and analyses the commands and calls the command specific action routines (see (b)). Most of its work is done by general subsystems independent of NFAS (see (e)).

(b) action routines

The command specific action routines correspond 1-to-1 to the NFAS commands. They are called from the routine flow control and return control to the same routine.

Id	Explanation
SE	SELECT, ENDSELECT
PK	PICK, ENDPICK
MO	MODIFY
CV	CONVERT
CH	CHANGE
MT	TRANSFORM
AU	TRANSFORM
NO	NORMALISE
DE	DELETE
SD	TREEDELETE
GR	MAKEMEMBERS
MM	MARKMEMBERS
MD	MAKEDISJUNCT
PL	PROTOCOLLTOLABEL

(c) type conversion routines

Id	Explanation
TA	Conic Arc → Parametric Spline Curve
TC	Constituents of Composite Curve → Independent Elements
TD	Copious Data → Lines/Points
TF	Subfigure Instance → Independent Elements
TG	Associative Instance/Group → Subfigure Instance/Definition
TO	Surface of Revolution → Circular Arcs etc.
TP	Plane → Ruled Surface etc.
TR	Ruled Surface → Copious Data/Lines etc.
TS	Parametric Spline Surface → Parametric Spline Curves
TT	Tabulated Cylinder → Ruled Surface

(d) NFAS specific subsystems and subroutines

Id	Explanation
AS	General subroutines for action routines
SL	Selection conditions
EL	Entity-Label-Protocol
ID	ID-Management
IC	NFAS conversions and checks

(e) general subsystems and subroutines providing services which can be used independently of NFAS.

Id	Explanation
PC	Procedure control
SA	Syntax analysis
MA	Memory allocation
SC	Sequencing control
LG	Logfile handling
ER	Error handling
ST	String handling
SY	System subroutines
IF	IGES file interface

B.1.2   Neutral File Call Interface

The IGES Call Interface is an independent software tool which can be used for further software development on the basis of the IGES format. It has not been developed within the CAD*I project and is only briefly considered in this chapter.

The IGES Call Interface is divided into user routines and system routines. The user routines communicate with application programs and the system routines communicate with each other by subroutine calls, function calls and common blocks.

The user routines may be separated into groups of modules by their different tasks like:

- Creation of entities,

- Reading of entities,

- Modification of entities,

- Deletion of entities,

- Modification of structures and

- Navigation in structures.

The system routines may be separated into the groups:

- System manager

  to control the admissibility of a call to the user routine,

- System administrator

  to administrate internal control parameters and tables and

- System accessor

  to execute the read and write access to the IGES file.

B.1.3 Programming language and operating systems

To avoid problems with the portability of the software all modules are written in FORTRAN77 complying with the ANSI X3.9-1978 standard. In addition the INCLUDE statement is used. Other system specific extensions are gathered in a documented group of modules only.

Up to now implementations of NFAS are running on the following operating systems:

* IBM VM/CMS,

* VAX VMS and

* MS-DOS with reduced functionality and restrictions in model size.

A UNIX based implementation is under development.

B.2   The performance of NFAS

B.2.1   The NFAS command language

NFAS is controlled by a simple procedural language which is interpreted during execution of the program. Although an interpreter is used, an enormous efficiency has been achieved by special treatment of repetitive commands. For the design of the command language the following points have been considered:

- an easy command syntax,

- functionality of commands similar to interactive CAD operations,

- naming of the commands with respect to their task and

- upward compatibility for extensions.

However, the NFAS command language is not intended to be a programming language. It is strictly limited to the functionality which is needed. All functions of higher complexity are defined as single self contained commands. The syntax of the commands is similar to DCL (DEC command language) and allows any abbreviation as far as the significance of a command is not affected.

At the moment NFAS offers 16 different commands which can be classified as shown below:

* changing of presentation attributes

- MODIFY command

* changing of entity parameters

- CHANGE command

* exploitation of structures

- MAKEDISJUNCT command

* generation of structures

- MARKEMEMBERS command

- MAKEMEMBERS command

*   entity type conversions and transformations

-   CONVERT command

-   NORMALISE command

-   TRANSFORM command

-   AUTOROTATE command

*   deletion of entities

-   DELETE command

-   TREEDELETE command

*   selection of entities for manipulation

-   SELECT command

-   ENDSELECT command

-   PICK command

-   ENDPICK command

*   protocollation of manipulations

-   PROTOCOLLTOLABEL command

B.2.2   The functionality of NFAS

The functionality of NFAS will be enhanced step by step as it is required by new applications or new CAD systems which have to be connected to the BMW CAD/CAM system environment. This presupposes an open system architecture of NFAS which allows an easy link to new routines to the existing system (see Figure 50). This modular design provides flexibility in the system with the ability to provide a quick extension for any functionality required.

The functionality of NFAS may be classified by seven types of functions:

- modification of attributes

- modification of geometry

- modification of structure information

- conversion between geometry representations

- selection of subsets of a neutral file

- deletion of subsets of a neutral file

- insertion of information

MODIFICATION OF ATTRIBUTES

This function is supported by the current version of NFAS for the attributes line font, level, blank status, entity use, line weight, color, entity name.

The function can be used in two ways:

The first way is to define the attribute by a constant value. The second way is to define the attribute by the value of another attribute of the same entity, which allows a simple way of mapping the information of one attribute (which may not exist in the receiving CAD/CAM system) to another attribute. The values are checked for conformity in both ways, with the allowed value ranges defined in the neutral file standard.

MODIFICATION OF GEOMETRY AND TEXT

'Modification of geometry' means either every change to the parameters of an entity or change of the location and orientation of the entity in space by a modification of its transformation matrix.

The transformation and the rotation of entities is one function. By applying this function not only the subset of entities selected will be transformed and rotated, but also all other entities which are physically dependent on them. One special form of rotation is the auto-rotation of entities from an unknown source plane into a defined location. In this case the user has to be sure that the chosen set of entities is coplanar.

It is intended to introduce additional geometry modification functions, in particular a function for the projection of 3-D geometry into an expressive form of 2-D geometry, e. g. the projection of 3-D geometry into 2-D space making use of the viewing information stored within the neutral file.

The modification of the text entity (general note) allows :

- the change of character height,

- the change of text ratio,

- the change of slant angle and furthermore

- the replacement or modification, if necessary, of text strings e.g. in the case of a change of length of a string the text box is changed to be consistent with the modified text.

MODIFICATION OF STRUCTURE INFORMATION

As different CAD/CAM systems use different methods of structuring, this group of functions is very important in order to improve the transferable data volume.

One function does the conversion between detail or subfigure structures and simple group structures in both directions. To go from subfigure to group, a logical group of the entities referenced by the subfigure definition entity must be created and connected with a transformation matrix to put it into the correct position. If there is more than one subfigure instance the entities have to be copied. Scale factors have not been considered up to now. From a group to a subfigure instance representation, a subfigure definition entity has to be generated containing the entities referring to the group entity. In addition, a subfigure instance entity which transforms the entities to the correct position is needed.

A lot of CAD/CAM systems do not work with groups. Therefore another function was implemented to 'disjunct' groups by copying common elements.

CONVERSION BETWEEN GEOMETRY REPRESENTATIONS

This group of functions allows the conversion between different representations of information in CAD/CAM systems. The only function implemented up to now is the conversions of CAD surface models based on the parametric spline surface entity to a wireframe representation based on parametric spline curves, with a possibility of defining the number of curves in u- and v-direction which are to be created.

SELECTION OF SUBSETS OF A NEUTRAL FILE

Concerning this group of functions two forms of selection have to be distinguished. The first form is a simple selection of a subset out of the whole model according to the selection conditions given. The second form is a selection including not only the entities complying with the selection conditions but also the entities which are dependent on these. Special forms are the 'picking' of entities related to a certain VIEW entity or a certain TRANSFORMATION MATRIX entity. The option PURGE added to the pick command deletes everything in the model which is not within the actual set of entities selected by the pick command. In both cases all following commands are executed on the selected set of entities until the ENDSELECT or ENDPICK command closes the loop.

To select a special set of entities, conditions containing entity types, attributes and structure relations have to be set up and may be combined using boolean operators. The SELECT command may be nested once.

DELETION OF SUBSETS OF A NEUTRAL FILE

The DELETE function may be applied to a set of entities identified by the SELECT command. In this case only the selected entities and the entities physically dependent on them are deleted. If all to the selected set of entities and dependent entities are to be deleted, including logically dependent child entities, the command TREEDELETE must be used.

INSERTION OF INFORMATION

The possibility of transferring product model structures indicated by the use of presentation attributes by grouping entities is a helpful function for systems unable to treat some types of presentation attributes like color, line weight, line font and so on. The implemented function MAKEMEMBERS to generate groups allows the grouping of a selected subset or the automatic grouping of elements which have the same value for a specified attribute type.

Another way of inserting information is to mark all physically or logically dependent entities of a selected subset by modification of an attribute using the MARKEMEMBER command. It is possible to increment this attribute value for each element of the selected subset. This mark may be used later for another selection.

B.3   Implementation of NFAS

As previously described, NFAS is used as one important tool within data transfer process using neutral file formats. At BMW the CAD system CATIA is the basic system within the product design phase. However in the whole product development and manufacturing planning a great number of other more specialist CAx systems are used and have therefore to communicate with other systems. To optimise the data exchange process the system NFAS is used and made available in the environment of the CAD systems. Figure 51 shows the actual placement of NFAS within the BMW CAx landscape.

As shown in the figure NFAS is available on the mainframe computers and included

NEUTRAL FILE ADAPTING SYSTEM

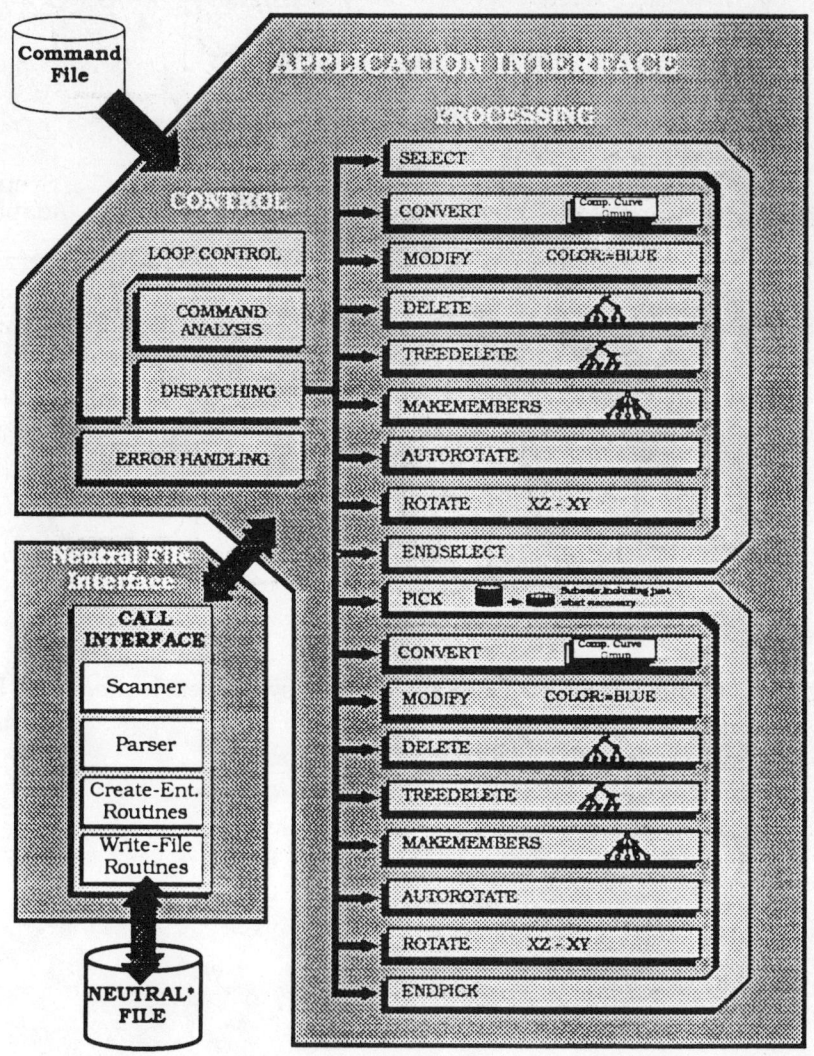

Figure 50. The open system architecture of NFAS

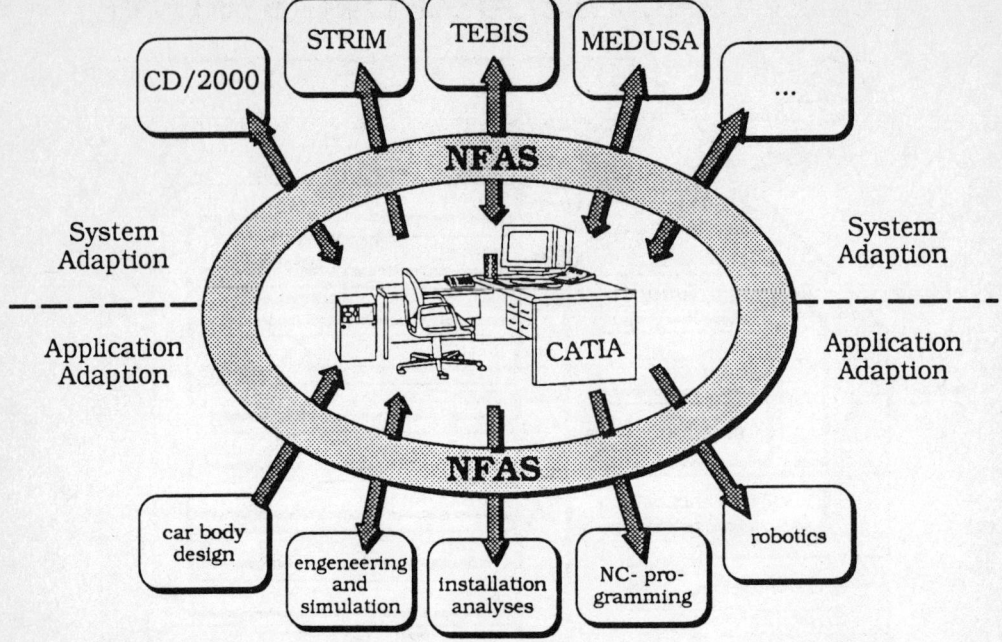

Figure 51. Placement of NFAS in the CAx system environment of BMW

within the environment of the following CAx systems:

* ICEM DDN

* CATIA

* CAEDS

* STRIM 100

* CABLOS (MEDUSA 2D)

* CADCPL (EXAPT)

List of Tables